SpringerBriefs in Computer Science

T0183694

For further volumes:
http://www.springer.com/series/10028

Hong Wen

Physical Layer Approaches for Securing Wireless Communication Systems

Springer

Hong Wen
University of Electronic Science
    and Technology of China
Chengdu
Sichuan
People's Republic of China

ISSN 2191-5768   ISSN 2191-5776 (electronic)
ISBN 978-1-4614-6509-6 ISBN 978-1-4614-6510-2 (eBook)
DOI 10.1007/978-1-4614-6510-2
Springer New York Heidelberg Dordrecht London

Library of Congress Control Number: 2013930686

Printed on acid-free paper

Springer is part of Springer Science+Business Media (www.springer.com)

Preface

Along with the rapid development of wideband wireless communication networks, wireless security has become a critical concern. Traditionally, in current wired or wireless communication networks, the issue of security is viewed as a whole independent feature addressed above the physical layer and cryptographic protocols are widely used to guarantee the security of the network. And the cryptographic protocols are designed assuming the physical layer has already been established and is error free. However, this assumption is usually impractical in the case of wireless communication. Compared with wireline networks, wireless networks lack a physical boundary due to the broadcasting nature of wireless transmissions. Any receivers nearby can hear the transmissions, and can potentially listen/analyze the transmitted signals, or conduct jamming. This unique physical-layer (PHY) weakness has motivated innovative PHY security designs in addition to, and integrated with, the traditional data encryption approaches.

One of the fundamental issues for PHY security is defined as information theoretic security, i.e., the adversary's received signal gives no more information for eavesdropping than legitimate receiver. The information-theoretic secrecy was first introduced by Shannon. "Perfect Secrecy" is defined by requiring of a system that after a cryptogram is intercepted by the enemy the a posteriori probabilities of this cryptogram representing various messages be identically the same as the a priori probabilities of the same messages before the interception. It is shown that perfect secrecy is possible by two approaches. First, perfect secrecy is possible but requires, if the number of messages is finite, the same number of possible keys. Second, perfect information theoretic secrecy requires that the signal Z received by the eavesdropper does not provide any additional information about the transmitted message W. Therefore, the build-in security of PHY security is also defined as: no secret keys are required before transmission.

In information-theoretic secrecy, channel noise plays a role of randomness resource. This book gives a review of the previous outstanding works of PHY security, and then provides the recent achievements on the confidentiality and authentication for wireless communications systems by channel identification. The first chapter introduces the PHY confidentiality and authentication concept. Chapter 2 introduces a practical approach to build unconditional confidentiality for wireless communication security by feedback and error correcting code. In wireless channels, multiple antennas can increase system robustness against fading, and also transmission rates, as well as providing valid ways to realize information-theoretic secrecy. A framework of PHY security based on space time block code (STBC) MIMO system is introduced in Chap. 3. Innovative cross-layer security designs with both PHY security and upper-layer traditional security techniques are desirable for wireless networks. In this chapter, we also present a scheme that combines cryptographic techniques implemented in higher layers with the physical layer security approach using redundant antennas of MIMO systems to provide stronger security for wireless networks. The channel responses between communication peers have been explored as a form of fingerprint with spatial and temporal uniqueness. Chapters 4 and 5 fulfill this idea and develop a new lightweight method of channel identification for Sybil attack and node clone detection in wireless sensor networks (WSNs).

Acknowledgments

The author wishes to acknowledge the partial support of the NSFC (Project No. 61032003, 61071100 and 61271172), NCET (Project No. NCET-09-0266), and the National Important Special Fund Project of China (Project No. 2011ZX03002-005-03). The author is thankful to Master candidates: Mr. Guo Chao Liu and Ms. Xiao Cheng for their contribution to editing work and figure reproduction. The author would also like to thank Prof. Bin Wu for his careful proof reading.

Very special thanks to Prof. Sherman Shen who made this book possible.

Contents

Chapter 1
Introduction for PHY Security

1.1 Introduction

Wireless networks lack a physical boundary due to the broadcasting nature of wireless transmissions. They are open to outside intrusions without the need of physical connections. The wireless security has become a critical concern in the physical layer. Physical-layer (PHY) security techniques, which are based on the Shannon secrecy model [1] are effective in resolving the boundary, efficiency and link reliability issues. In addition, the security in classical cryptography system is based on unproven assumptions regarding the hardness of certain computational tasks. Therefore, systems are insecure if assumptions are wrong or if efficient attacks are developed.

In the classic model of a cryptosystem introduced by Shannon [1], both the sender and the intended receiver share a common secret key, which is unknown to the wiretapper, and use this key to encrypt and decrypt the message M. Shannon considered a scenario where both the intended receiver and the wiretapper have direct access to the transmitted signal. If the signal received by the wiretapper is Z, perfect security was defined as $I(M, Z) = 0$, i.e. M and Z are statistically independent.

The built-in security of the physical-layer is defined as the physical-layer transmissions which guarantee low-probability-of-interception (LPI) based on transmission properties such as modulations, signals and channels, without resorting to source data encryption. No secret keys are required before transmissions. An alternative notion of communication with perfect secrecy was introduced by Wyner [2] and later by Csiszar and Korner [3], who developed the concept of the wiretap channel for wired links. Wyner proves that the perfect secrecy can be realized if and only if the eavesdropper's channel is noisier than the legitimate receiver's channel.

Based on Wyner's model, the authors use multiple antennas system to construct perfect secret system in [4–6]. Several authors presented the LDPC [10, 11] as secret code in wiretap channel. In [13], the authors exploit user cooperation in facilitating the transmission of confidential messages from the source to the

H. Wen, *Physical Layer Approaches for Securing Wireless Communication Systems*, SpringerBriefs in Computer Science, DOI: 10.1007/978-1-4614-6510-2_1, © The Author(s) 2013

destination. Tekin and Yener [14] consider multiple-access two way wiretap channel. Lai et al. [15] proposed the feedback approach to build the wiretap channel. Hero [4] and Koorapaty et al. [29] presented an information security approach which uses channel state information (CSI) as the secret key in multiple-input multiple-output (MIMO) links. Based on these principles, Zhang and Dai [30] designed a class of unitary space–time ultra-wideband (UWB) signals to achieve the perfect communication secrecy.

In this book, we are going to introduce two approaches to build perfect security from feedback and MIMO technologies, which mainly utilize the unique channel characteristics as the secret key. There are no computational restrictions to be placed on the eavesdropper in physical-layer security systems.

The unique channel characteristics of wireless channel can also be used as the PHY authentication. The main idea of is based on a generalized channel response with both spatial and temporal variability, and considers correlations in the time, frequency and spatial domains. It has been envisioned that the physical layer security mechanisms are promising to achieve fast authentication and low overhead in the future communication networks, particularly in a highly dynamic networking environment like VANETs and Ad Hoc Wireless Sensor Networks (AWSN). However, the PHY authentication has to depend on the upper layer traditional authentication for initial identity. Therefore, we introduce the cross-layer authentication scheme by combining the PHY channel information identity with upper layer authentication.

The idea of physical layer authentication is also a good candidate for identifying some malicious attack in energy limited scenarios because of its fast and lightweight performance. The node clone attack and Sybil attack are two awfully harmful attacks for Wireless Sensor Networks (WSNs). WSNs is an interconnected system of a large set of physically small, low cost, low power sensors that provide ubiquitous sensing and computing capabilities.

Three general ways is proposed to detect clone nodes as following. The idea about location-based keys to thwart and defend against the node clone is presented in [79, 80]. References [81–84] introduced the key identity validation method and the identity of entity validation approach, which employed cryptographic related method to prevent node clone attacks. Sheng et al. [85] presented the radio resource testing approach as a defense against node replication attack, which is based on the assumption that a radio can not send and receive simultaneously on more than one channel. However, methods based on the accurate geographic location or cryptographic-based schemes are difficult to be implemented into WSNs since the hardware of WSNs nodes is inefficient. PHY identification of the node clone and Sybil attack is a good solution for WSNs.

In short, although traditional approaches to network security are important to secure wireless systems, they cannot protect against the full range of threats facing wireless networks, nor provide a complete toolbox to protect wireless networks. The objective of this book is to highlight the importance of new paradigms for securing wireless systems that take advantage of wireless-specific properties to thwart security threats.

1.2 Outline

This book focuses on PHY securing wireless communication and aims to introduce the basic concept about PHY confidentiality and authentication and present some recent novel research results in this area. Following is an overview of this book.

Chapter 2: Unconditional security communication is defined as no secret keys sharing before transmissions while no information can leak to eavesdroppers. According to Wyner's model, two steps should be taken for realizing unconditional security communication. In this chapter, firstly a multi-round interactive communication scheme is introduced and the information stream is hidden in the additional noise impairing the eavesdropper by using the random sequence that only be known between the legitimate partners. As a result, the eavesdropper can receive all these signals, but his received signals include an additional noisy that produced by his own channel. The wire tap channel is built without pre-shared a secret key. Secondly, the security codes are put on top of the interactive communication model, aiming to achieve an error-free legitimate channel while keeping the eavesdropper from any useful information (i.e., with an error probability of 0.5). By integrating a multi-round two-way communication model with a security code, an unconditional security model is built such that the wiretapper is subject to an error probability close to 0.5 while the main channel is almost error-free. Instead of raising any unpractical assumption on the wiretapper's channels, the proposed approach in this chapter is applicable in any scenario with known noisy channels from the sender to the intended receiver and to the wiretapper, respectively. Thus, the proposed approach yields practical usage in many circumstances, such as on achieving confidentiality in a symmetric cryptographic system for key exchange, distribution, and message confidentiality.

Chapter 3: Multiple-input multiple-output (MIMO) links can provide rich channel characteristics, which are resources for building PHY confidentiality. However, the PHY security is an average-information measure. The system can be designed and tuned for a specific level of security e.g., with very high probability a block is secure, but it may not be able to guarantee security with probability 1. Any deployment of a physical-layer security protocol in a classical system would be part of a "layered security" solution where security is provided at a number of different layers, each with a specific goal in mind. Innovative cross-layer security designs with both physical-layer security and upper-layer traditional security techniques are desirable for wireless networks. In this chapter, we propose a different scheme with a cross-layer approach to enhance the security of wireless networks for wireless environments. We combine cryptographic techniques implemented in the higher layer with the physical layer security scheme using redundant antennas of MIMO systems to provide stronger security for wireless networks. By introducing a distort signal set instead of an orthogonal codeset for wireless networks based on space time block code (STBC) system, the transmitter randomly flip-flops between the distort signal set and the orthogonal code set for confusing the attacker. An upper-layer pseudorandom sequence will be employed to control the flip-flops process. In this approach the physical-layer can utilize upper-layer encryption techniques for security, while physical-layer security techniques can also assist the security design in the upper-layer.

Chapter 4: Authentication of the broadcast messages is an effective approach to countermeasure most of the possible attacks, by which the intended receivers can make sure that the received data is originated from the expected source. However, Public Key Infrastructure (PKI) based digital signature authentication techniques are too resource consumption to perform in wireless sensor networks. With physical layer authentication, a channel response is extracted and used to estimate the channel information before the signals received. It has been envisioned that the physical layer security mechanisms are promising to achieve fast authentication and low overhead in future communication networks, particularly in a highly dynamic networking environment like VANETs and Ad Hoc Wireless Sensor Networks (AWSN). Therefore in this chapter, we explore the potential possibility of using physical-layer channel responses as authenticators between each communication pair, and propose two novel message authentication schemes. Under IEEE802.11p physical layer format and IEEE802.11a physical layer packets, the preamble of each Orthogonal Frequency-Division Multiplexing (OFDM) packet can be used for channel estimation, where no extra bandwidth or transmission power will be consumed. By combining the upper layer authentication scheme with the PHY authentication scheme, a cross-layer PHY authentication framework called Physical-layer Assisted Authentication (PAA) is built for achieving fast and light-weighted message authentication in wireless networks.

Chapter 5: In this chapter, we follow the idea in the previous chapter to introduce a new lightweight method for Sybil attack and node clone detection in WSNs, which is based on Channel Identification (CI). A Sybil node or a clone node is distinguished by the channel response which is extracted and used to estimate the channel information between two nodes. No geographic location detection and complexity cryptographic computation is needed. Therefore, these approaches incur much less data transmission and light overhead. Finally, by employing Wireless InSite (WI) tools, NS-2 and MATLAB software simulations are performed to validate these schemes under different networks.

Chapter 2
Unconditional Security Wireless Communication

Wyner [2], Csiszar and Korner [3] developed the concept of the wiretap channel for wired links. In a wiretap channel, the eavesdropper is assumed to receive messages transmitted by the sender over a channel that is noisier than the legitimate receiver's channel. Under this condition, it is possible to establish perfectly secure source–destination link without relying on secret keys. Unfortunately, it may often be impossible to guarantee that the adversary's channel is noisier than the one of the legitimate partner. In a general communication model, it is possible that the received signal of the intended receiver is worse than the received signal of the eavesdropper. Wyner proved that the transmitter could send information to the legitimate receiver in virtually unconditional secrecy without sharing a secret key with the legitimate receiver if the eavesdropper's channel is a degraded version of the main channel.

In the 1970s and 1980s, the impact of these works was limited, partly because practical wiretap codes were not available, but mostly because a strictly positive secrecy capacity in the classical wiretap channel setup requires the legitimate sender and receiver to have some advantage over the wiretapper in terms of channel quality. In Wyner's model, building unconditional security communication generally takes the following two steps. Firstly, a practical wiretap channel is built by initiating advantages for the legitimate communication peers in terms of channel quality against the eavesdroppers; while the second step is to achieve the unconditional secure communication by security codes. In this chapter, we first present a different approach to build the wiretap channel is presented, and then intorduce the Wyner coset codes.

In this chapter, firstly we utilize the feedback and low density parity check (LDPC) codes to build the wiretap channel I [2] in which the eavesdropper sees a binary symmetric channel (BSC) with error probability p and the main channel is error free. In our model, we consider the situation that an external eavesdropper can receive the signals through the main channel and the feedback channel. Firstly, we developed the Maurer's idea [7] and used a powerful interactive communication to build an unconditionally-secure communication model in which the eavesdropper's channel is noisier than the legitimate partner. Then we utilize the

H. Wen, *Physical Layer Approaches for Securing Wireless Communication Systems*, SpringerBriefs in Computer Science, DOI: 10.1007/978-1-4614-6510-2_2, © The Author(s) 2013

threshold property of LDPC codes [19] to correct the error of the main channel and remain the error of eavesdropper's channel. In the second step, we developed the security codes on top of the interactive communication model in [16], aiming to achieve an error-free legitimate channel while keeping the eavesdropper from any useful information (i.e., with an error probability of 0.5). By integrating a multi-round two-way communication model with a security code, an unconditional security model is built such that the wiretapper is subject to an error probability close to 0.5 while the main channel is almost error-free.

2.1 Perfect Security Model

Consider a communication system consisting of a source, a destination and an eavesdropper as shown in Fig. 2.1. The source produces a message $S = [s_1 \quad s_2 \cdots s_m]$, and encodes this message as a vector $X = [x_1 \quad x_2 \quad \cdots \quad x_n]$. This vector is transmitted through a communication main channel and received as $Y = [y_1 \quad y_2 \cdots y_n]$ by the destination. The eavesdropper has access to t $(t < n)$ symbols of X through a communication wiretap channel and we denote the received vector of an eavesdropper as $Z = [z_1 \quad z_2 \quad \cdots \quad z_n]$. The destination and eavesdropper can decode information as $\hat{S} = [\hat{s}_1 \quad \hat{s}_2 \quad \cdots \quad \hat{s}_m]$ and $\tilde{S} = [\tilde{s}_1 \quad \tilde{s}_2 \quad \cdots \quad \tilde{s}_m]$ from Y and Z, respectively. If the error rate of the destination and eavesdropper are:

$$P_e^m = \frac{1}{m} \sum_{i=1}^{m} \Pr(\hat{s}_i \neq s_i) \tag{2.1}$$

$$P_e^w = \frac{1}{m} \sum_{i=1}^{m} \Pr(\tilde{s}_i \neq s_i), \tag{2.2}$$

The two conditions for security communication are:

$$P_e^m \to 0 \tag{2.3}$$

$$P_e^w \to 0.5 \tag{2.4}$$

Condition (2.3) is called reliability condition which implies that X must be determinate by S. Condition (2.4) is called security condition such that an

Fig. 2.1 Wyner's perfect security model

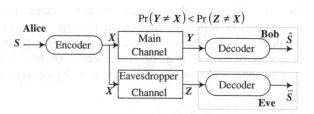

eavesdropper can not get any useful information from t intercepted symbols. An alternately expression can be as following: the equivocation of the system is

$$\Delta = H(S^m | Z^n) \tag{2.5}$$

which means that the eavesdropper's remaining uncertainty about the source vector is at least Δ. When $\Delta = K$, the eavesdropper obtains on information about the source, and the system obtained perfect secrecy, which means that the transmitter could send information to the legitimate receiver in virtually unconditional secrecy without sharing a secret key with the legitimate receiver. Such model is called perfect security model.

2.2 Powerful Multi-round Feedback for Build Wiretap Channel Model

2.2.1 Two Way Communication for Build-in Wiretap Channel

Wyner proved that the transmitter could send information to the legitimate receiver in virtually perfect secrecy without sharing a secret key with the legitimate receiver if the eavesdropper's channel is a degraded version of the main channel. The assumption that the adversary only receives a degraded version of the legitimate receiver's information is unrealistic in general. But this assumption is impractical. Sometimes the adversary can have a better channel than that of legitimate user. In [7], two way communications are developed to realize the practical adversary degraded channel building.

We introduce the two way communications approach as following. Let $E = \{e_0, e_1, \cdots, e_{n-1}\}$ and $EA = \{ea_0, ea_1, \cdots, ea_{n-1}\}$ denote the error vectors of the Alice's and the eavesdropper's channel, respectively. The received signals of Alice and the eavesdropper are

$$
\begin{aligned}
T &= \{t_0, t_1, \cdots, t_{n-1}\}, t_i = q_i \oplus e_i \\
TE &= \{te_0, te_1, \cdots, te_{n-1}\}, te_i = q_i \oplus ea_i
\end{aligned}
\tag{2.6}
$$

where $P_r(e_i = 1) = \alpha$ and $P_r(ea_i = 1) = \beta$. Then Alice use the received signal T to calculate

$$U = \{u_0, u_1, \cdots, u_{n-1}\}, u_i = t_i \oplus m_i \tag{2.7}$$

and encode U such that

$$W = \phi(U) \tag{2.8}$$

where ϕ is the encoder function. Alice sends W over the channel. Alice and the eavesdropper receive the noise version of W as W' and decode W' as

$$\tilde{U} = \psi(W') \tag{2.9}$$

where ψ is the decoder function. We assume the decoding error probability $P_r(\tilde{U} \neq U) \to 0$. Bob and the eavesdropper received the U with almost error free. Bob knows the random sequence Q, so he can add wise Q to U as

$$Y = U \oplus Q = M \oplus E \tag{2.10}$$

where $Y = \{y_0, y_1, \cdots, y_{n-1}\}$. The eavesdropper only knows TE that is the noise version of Q and he only can add wise Eq. (2.6) to U as:

$$Z = U \oplus TE = M \oplus E \oplus EA \tag{2.11}$$

where $z = \{z_0, z_1, \cdots, z_{n-1}\}$. Comparing Eq. (2.10) with Eq. (2.11), EA becomes the extra noise. Therefore, after two way communication in Fig. 2.2, the direction of the main channel is inverted when the eavesdropper initially has a better channel.

Lemma 2.1 *After two way communication, the error probability of main channel is α and the error probability of eaveasdropper's channel is $\alpha + \beta - 2\alpha\beta$.*

Proof Since

$$Pr(y_i \neq m_i) = Pr(e_i = 1),$$

and

$$Pr(z_i \neq m_i) = Pr(e_i = 1) \cdot Pr(ea_i = 0) + Pr(e_i = 0) \cdot Pr(ea_i = 1)$$

Thus $r(z_i \neq m_i) = \alpha + \beta - 2\alpha\beta$.

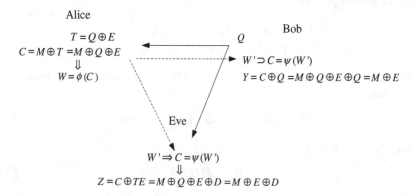

Fig. 2.2 Two way communication

Because $\alpha \leq 0.5$ and $\beta \leq 0.5$, so we have $\alpha \leq \alpha + \beta - 2\alpha\beta$, where the equality holds for $\alpha = 0.5$ or $\beta = 0.5$.

2.2.2 Multi-round Feedback for Build Wiretap Channel Model

In [7], the secrecy capacity Cs is defined as the maximum rate at which a transmitter can reliably send information to an intended receiver such that the rate at which the attacker obtains this information is arbitrarily small. If the channel from the transmitter to the intended receiver and the channel from the transmitter to the eavesdropper have different bit error probabilities (BER) δ and ε, respectively, the secret capacity Cs is [7].

$$C_s = \begin{cases} h(\delta) - h(\varepsilon), & \text{if } \delta > \varepsilon \\ 0, & \text{otherwise} \end{cases} \tag{2.12}$$

where h denotes the binary entropy function defined by

$$h(p) = -p\log_2 p - (1-p)\log_2(1-p)$$

After two way communication, the main channel has advantage over the eavesdropper's channel. One of his attacking methods to the wiretap channel is eliminating the secrecy capacity which means trying to let $\delta \leq \varepsilon$. When β is very small, from Eq. (2.12) we know that the secrecy capacity Cs is very small. The secret lever of system is weak. By several rounds of two way communication or several parallel channel feedbacks, the advantage of the main channel can be increased. Our scheme to build wiretap channel I is presented as following. To transmit k-bit messages M, we first select a (n, k) linear binary code C such that

$$C = \chi(M) \tag{2.13}$$

where χ is the encoder function which maps the k bits message M into a n bits codeword C. By randomly choosing $C_0, C_1, C_2, \cdots, C_{t-2}$, where $C_i = (c_i^0, c_i^1, c_i^2, \cdots, c_i^{n-1})$, $0 \leq i \leq t-2$, we can calculate the vector

$$C_{t-1} = C_0 \oplus C_1 \oplus C_2 \oplus \cdots \oplus C_{t-2} \oplus C \tag{2.14}$$

Firstly, Bob sends t random sequence $Q_i = (q_i^0, q_i^1, \cdots, q_i^{n-1})$, $i = 0, 1, 2, \cdots$ $t-1$ to Alice by the t independent parallel channels or a channel in different time slots. Let $E_i = (e_i^0, e_i^1, \cdots, e_i^{n-1})$ and $EA_i = (ea_i^0, ea_i^1, \cdots, ea_i^{n-1})$ denote the error vectors of the Alice's and the eavesdropper's channel corresponding to the transmitting the random sequence Q_i respectively. The received signals of Alice and the eavesdropper are T_i and E_i. Then Alice uses the received signal calculate $U_i = C_i \oplus T_i$ according to Eq. (2.7) and encode to get W_i according to Eq. (2.8). Alice sends W_i over the channel. Alice and the eavesdropper receive the noise version of W_i as W'_i and decode W'_i according to Eq. (2.9). From Eqs. (2.10) and (2.11), Bob and the eavesdropper can get $Y_i = C_i \oplus E_i$ and $Z_i = C_i \oplus E_i \oplus EA_i$. Here we consider the discrete memoryless channel (DMC). We assume that the t words $C_0, C_1, C_2, \cdots, C_{t-1}$ and t random sequence are transmitted from the t independent parallel channels or a channel in t different time slots, respectively. In each time slot the transmitting signals are independent. Our scheme is shown in Fig. 2.3.

We sum the $Y_i, i = 0, 1, 2, \cdots, t-1$ and $Z_i, i = 0, 1, 2, \cdots, t-1$ respectively as

$$Y = \sum_{i=0}^{t-1} Y_i = \sum_{i=0}^{t-1} C_i \oplus \sum_{i=0}^{t-1} E_i \tag{2.15a}$$

$$Z = \sum_{i=0}^{t-1} Z_i = \sum_{i=0}^{t-1} C_i \oplus \sum_{i=0}^{t-1} E_i \oplus \sum_{i=0}^{t-1} EA_i \tag{2.15b}$$

According to Eq. (2.14), Eqs. (2.15a) and (2.15b) become:

$$Y = C \oplus \sum_{i=0}^{t-1} E_i$$

$$Z = C \oplus \sum_{i=0}^{t-1} E_i \oplus \sum_{i=0}^{t-1} EA_i \tag{2.16}$$

The term $\sum_{i=0}^{t-1} EA_i$ in Eq. (2.16) becomes the extra error. Therefore, what we want to do is to extract the correct information M from Y and keep the extra error

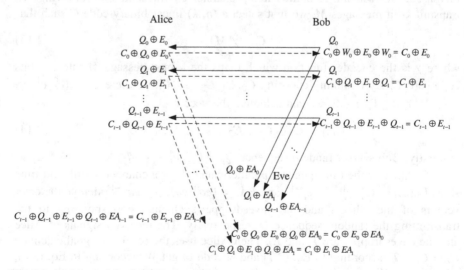

Fig. 2.3 Multi-round feedback communication

remaining in Z by decoding. Then we can get the wiretap channel model I [2] in which the main channel is almost error free and the eavesdropper's channel is noisy.

Lemma 2.2 *Let the error probability of E_i denote* $\Pr(e_j^i = 1) = \alpha_i$ *and the error probability of EA_i denote* $\Pr(ea_j^i = 1) = \beta_i$, *the error probability of Y in Eq. (2.15a) and Z in Eq. (2.15b) are:*

$$p(Y) = \sum_{i=0}^{t-1} \alpha_i - 2 \sum_{\substack{i_1,i_2=0 \\ i_1 > i_2}}^{t-1} \alpha_{i1}\alpha_{i2} + 4 \sum_{\substack{i_1,i_2,i_3=0 \\ i_1 > i_2 > i_3}}^{t-1} \alpha_{i1}\alpha_{i2}\alpha_{i_3} + \cdots + (-1)^t 2^t$$

$$\sum_{\substack{i_1,i_2,i_3=0 \\ i_1 > i_2 > i_3}}^{t-1} \alpha_{i1}\alpha_{i2}\alpha_{i_3}\cdots\alpha_{i_t}$$

$$(2.17)$$

$$P(Z) = P(Y) + P(ZEA) - 2P(Y)P(ZEA)$$

where

$$p(ZEA) = \sum_{i=0}^{t-1} \beta_i - 2 \sum_{\substack{i_1,i_2=0 \\ i_1 > i_2}}^{t-1} \beta_{i1}\beta_{i2}$$

$$+ 4 \sum_{\substack{i_1,i_2,i_3=0 \\ i_1 > i_2 > i_3}}^{t-1} \beta_{i1}\beta_{i2}\beta_{i_3} + \cdots + (-1)^t 2^t \sum_{\substack{i_1,i_2,i_3=0 \\ i_1 > i_2 > i_3}}^{t-1} \beta_{i1}\beta_{i2}\beta_{i_3}\cdots\beta_{i_t}$$

The proof is derived from lemma 2.1 directly.

By interactive communication we can get an unconditionally-secure communication model in which the legitimate partner can realize the secret information transmitting without pre-shared secret key even if the eavesdropper have better channel at the beginning. The secret strength of the wiretap channel partly depends on the secret capacity. In our scheme, the choosing a larger number of interactive communication round (parameter t) can lead to larger the secret capacity. The feedback sequences are chosen randomly and error vectors caused by the channel noise are random. The feedback signals from the destination plays the role of a private key. Therefore, if one data frame is broken, the other data frames still keep secret.

2.3 Performance with LDPC Codes

Turbo codes [16] and LDPC codes [17, 18] have already proved their excellent performance for error correction therefore both are good candidates for our scheme. This book focuses on LDPC codes although we believe that turbo-codes or any other strong channel codes would yield similar results. But the long codeword length is necessary, which can let the exhaust attacking complexity of the attacker become high.

2.3.1 The Threshold Property of LDPC Codes

LDPC code has been shown to provide excellent decoding performance that can approach the Shannon limit in some case. The LDPC code exhibits a threshold phenomenon with certain decoding method, which determines the asymptotic (in the codeword length) behavior of the ensemble of code: roughly speaking, for a code chosen randomly from the ensemble, with high probability decoding will be successful if transmission takes place below this threshold, and the error probability will stay above a fixed constant if transmission takes place above this threshold. This property can let us correct the error of the main channel and remain the error of eavesdropper's channel at the some time. Because we consider the BSC channel, we use Bit-Flip (BF) iterative decoding method, which was devised by Gallagher in the early 1960s [17].

Firstly, we discuss the upper bound of threshold that can be decoded correctly. We consider the regular LDPC codes, define a (n, d_l, d_r) parity-check matrix as a matrix of n columns that has d_l ones in each column, d_r ones in each row, and zeros elsewhere. Let p_0 denote the crossover probability of the binary-symmetric channel. It was shown in [19] that the expected number of errors in the ith iteration is given by the recursion:

$$a_i = p_0 - p_0 f^+(a_{i-1}) + (1 - p_0)f^-(a_{i-1}) \tag{2.18}$$

where $f^+(x) = \lambda\left(\frac{1+\rho(1-2x)}{2}\right)$, $f^-(x) = \lambda\left(\frac{1-\rho(1-2x)}{2}\right)$ and the degree distribution pair $(\lambda(x), \rho(x))$ are function of the form: $\lambda(x) = \sum_{j=2}^{\infty} \lambda_j x^{j-1}$, $\rho(x) = \sum_{j=2}^{\infty} \rho_j x^{j-1}$, where λ_j and ρ_j denote the fraction of ones in the parity-check matrix of the LDPC code which are in columns (rows) of weight j.

Definition 2.1 The threshold p_{up}^* is the supremum of all p_0 in $\left[1, \frac{1}{2}\right]$ such that as defined in (2.18) converges to zero as i tend to infinity.

Lemma 2.3 *Let τ denotes the smallest positive real root of the polynomial and* $p(x) = xf^+(x) + (x - 1)$
$f^-(x)$ and $\lambda_2\rho(1) < 1$ hold. Then

$$P_{up}^* \leq \min\left\{\frac{1 - \lambda_2\rho'(1)}{\lambda'(1)\rho'(1) - \lambda_2\rho'(1)}\right\} \tag{2.19}$$

where $\lambda'(x)$ and $\rho'(x)$ denote derivatives of $\lambda(x)$ and $\rho(x)$, respectively [19].

Table 2.1 The thresholds of some regular LDPC code ensembles

d_l	d_r	Rate	p_{up}^*	$p_{0.2}^*$	d_l	d_r	Rate	p_{up}^*	$p_{0.2}^*$
3	6	0.5	0.04	0.145	4	5	0.2	0.123	0.255
4	8	0.5	0.048	0.145	5	6	0.167	0.142	0.279
3	5	0.4	0.061	0.178	7	8	0.125	0.175	0.0306
4	6	0.333	0.067	0.205	9	10	0.1	0.197	0.0325
3	4	0.25	0.107	0.228					

Then we consider the threshold that can not be decoded correctly. We use the Shannon limit in the BSC as the threshold that the channel error can not be corrected by decoding.

Definition 2.2 The threshold p_{ep}^* is the infimum of all p_0 in $\left[1, \frac{1}{2}\right]$ such that the average error probability of the codeword is greater than a constant number $ep \leq 0.5$ as the number of iterative decoding tends to infinity.

Therefore, we hope $P(Y) < p_{up}^*$ and $P(Z) > p_{ep}^*$ after several rounds of inter-active communication, which means the main channel will be almost error free and the eavesdropper still remains ep error probability that can not be corrected at least in the same time after LDPC code decoding.

Table 2.1 includes the thresholds of some regular LDPC code ensembles. We let $ep = 0.2$ in Table 2.1. The results of Table 2.1 are also shown in the Fig. 2.4.

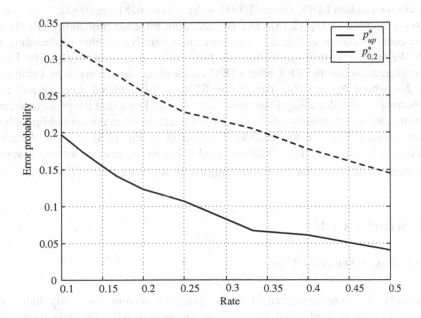

Fig. 2.4 The thresholds of some regular LDPC code ensemble

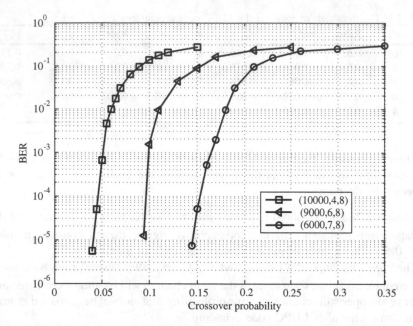

Fig. 2.5 The BER performances of LDPC codes

2.3.2 Some Performance Results

We use the random LDPC codes (10000, 4, 8), (9000, 6, 8) and (6000, 7, 8) as the encoder function in Eq. (2.13). The BF decoding bit error probability (BER) of these codes is shown in Fig. 2.5. The maximum iterative number of decoding is 200. We set $t = 1$ (two way communication) and $t = 2$, respectively. The BER after interaction and the BER after LDPC codes decoding are given in Table 2.2 and 2.3, where P_{ir} and P_{Eve} denote the BER of the intended receiver and the eavesdropper after decoding. From the results we can know that the positive secret capacity can be achieved by interaction communication when the eavesdropper has better channel. Because the encoder function φ in Eq. (2.8) and the decoder function ψ in Eq. (2.9) don't have essential effect to our results, we can assume that there are the powerful error correcting codes which can let $P_r(\widehat{U}_i \neq U_i) \rightarrow 0$.

2.4 Security Code

2.4.1 Coset Security Code

Obviously, the scheme described in the previous section goes only half-way to providing an unconditional security communication. After multiple rounds of

Table 2.2 The performance properties with $t = 1$

Crossover probability	$\alpha = 0.04$ $\beta = 0.04$	$\alpha = 0.08$ $\beta = 0.04$	$\alpha = 0.08$ $\beta = 0.08$	$\alpha = 0.15$ $\beta = 0.075$	$\alpha = 0.15$ $\beta = 0.15$
LDPC code	(1000, 4, 8)	(9000, 6, 8)	(9000, 6, 8)	(6000, 7, 8)	(6000, 7, 8)
BER after interaction	$P(Y) = 0.04$ $P(Z) = 0.0768$	$P(Y) = 0.08$ $P(Z) = 0.1136$	$P(Y) = 0.08$ $P(Z) = 0.1472$	$P(Y) = 0.015$ $P(Z) = 0.2138$	$P(Y) = 0.15$ $P(Z) = 0.255$
BER after decoding	$P_{ir} < 10^{-5}$ $P_{Eve} = 0.05$	$P_{ir} < 0.25 \times 10^{-5}$ $P_{Eve} = 0.011$	$P_{ir} = 1.25 \times 10^{-5}$ $P_{Eve} = 0.075$	$P_{ir} = 5.2 \times 10^{-5}$ $P_{Eve} = 0.095$	$P_{ir} = 5.2 \times 10^{-5}$ $P_{Eve} = 0.2$
Csin (7)	0.2862	0.0871	0.3841	0.4521	0.7211

Table 2.3 The performance properties with $t = 2$

Crossover probability	$\alpha_1 = \beta_2 = 0.02$ $\alpha_1 = \beta_2 = 0.02$	$\alpha_1 = \beta_2 = 0.04$ $\alpha_1 = \beta_2 = 0.02$	$\alpha_1 = \beta_2 = 0.04$ $\alpha_1 = \beta_2 = 0.04$	$\alpha_1 = \beta_2 = 0.08$ $\alpha_1 = \beta_2 = 0.04$	$\alpha_1 = \beta_2 = 0.08$ $\alpha_1 = \beta_2 = 0.08$
LDPC code	$(1000, 4, 8)$	$(9000, 6, 8)$	$(9000, 6, 8)$	$(6000, 7, 8)$	$(6000, 7, 8)$
BER after interaction	$P(Y) = 0.0392$ $P(Z) = 0.0753$	$P(Y) = 0.0768$ $P(Z) = 0.11$	$P(Y) = 0.0768$ $P(Z) = 0.1418$	$P(Y) = 0.01472$ $P(Z) = 0.201$	$P(Y) = 0.1472$ $P(Z) = 0.251$
BER after decoding	$P_{ir} < 10^{-5}$ $P_{Eve} = 0.048$	$P_{ir} < 1.25 \times 10^{-5}$ $P_{Eve} = 0.0093$	$P_{ir} = 1.25 \times 10^{-5}$ $P_{Eve} = 0.071$	$P_{ir} = 1.5 \times 10^{-5}$ $P_{Eve} = 0.055$	$P_{ir} = 1.5 \times 10^{-5}$ $P_{Eve} = 0.205$
C_s in (2.12)	0.2777	0.0759	0.3694	0.3070	0.7316

two-way communication, the two legitimate parties Alice and Bob are connected by a noiseless binary channel, and the wiretapper Eve receives the bits sent over the channel with some error probability $\varepsilon > 0$. Our object is to let error probability $\varepsilon = 0.5$. Formally, let $M = \{m_1, m_2, \cdots, m_k\}$ and $\hat{M} = \{\hat{m}_1, \hat{m}_2, \cdots, \hat{m}_k\}$ and $\hat{M}_E = \{\hat{m}_{E_1}, \hat{m}_{E_2}, \cdots, \hat{m}_{E_k}\}$ be vectors denoting Alice's message, Bob's decoded message and Eve's decoded message, respectively. Unconditional security is said to be achieved if the following relation holds:

$$\Pr(m_i \neq \hat{m}_i) = 0$$

$$\Pr(m_i \neq \hat{m}_{Ei}) = 0$$

(2.20)

To achieve this goal, we need to develop the security codes (SC) that can further degrade the wiretapper's information while without damnifying the legitimate users'. To the best of our knowledge, code construction and their relation to security, although explored in a few studies, are still subject to further research due to their deficiency for practical capacity approaching security code. Here we present some existing results and use it to complete the unconditional security communication scheme in the next section.

Following the special case considered by Wyner [2], we consider the coding method as following. To transmit k bits message $M^j = \{m_1^j, m_2^j, \cdots m_k^j\}$, $j = 1, 2, \cdots, 2^k$, a $(n, n - k)$ linear binary code C_e with coset $V = \{V^1, V^2, \cdots, V^{2^k}\}$ is chosen. Let each message M^j correspond to a coset $V^j = \{V_1^j, V_2^j, \cdots, V_{2^{n-k}}^j\}$ where n-tuple $V_i^j = w^j \oplus Ce^i$, $i = 2^{n-k}$, $j = 2^k$, w^j and $Ce^i \in \{0, 1\}$, Ce^i is codeword of $(n, n - k)$ linear binary code. We construct the encoder such that the encoder output $V_i^j \in \{0, 1\}^n$ is a randomly chosen member of the coset when the sending message is M^j. Therefore, message M^j and w^j corresponding the syndrome and error vector of the code C_e, respectively. Clearly, the decoder of the legitimate receiver can perfectly recover M^j from output V_i^j perfectly if the legitimate communication partners hold an error free channel. Now we turn to eavesdroppers who observe the noisy version $Ze^j \in \{0, 1\}^n$, which is the output of the BSC corresponding to the input V_i^j, We can state the security criterion to guarantee security of Alice's message M^j in the following lemma.

Lemma 2.4 *The average bit error rate* (BER) *of the recovering message M^j from Ze^j equals to 0.5, i.e.,* $\Pr(m_i \neq \hat{m}_{Ei}) = 0.5$, *when the received noisy version Ze^j has the equal probability to fall into one of cosets V, i.e.,,* $\Pr(Ze^j \neq V^j) = 2^{-k}$, *for $j = 1, 2, \cdots, 2^k$.*

Proof We have $\Pr(Ze^j \in V^j) = 2^{-k}$, which means that there is probability 2^{-k} to take arbitrary k bits vector from entire space $\{0, 1\}^k$. Let $A_{t,j}$ denote the number of k dimension vectors which have a distance t with vector M^j and

$\Pr\left(m_i^j \neq \hat{m}_{Ei}^j | Ze^j\right) = 2^{-k}$ denote the average BER of the recovered message M^j, we have:

$$\Pr\left(m_i \neq \hat{m}_{Ei} | Ze^j\right) = \left(\Pr(Ze^j \in V^j)\sum_{t=0}^{k}(A_{t,j}.t)\right)\Big/ k$$

$$= \left(2^{-k}\sum_{t=0}^{k}\binom{k}{t}\cdot t\right)\Big/ k$$

$$= \frac{1}{2}$$

Now the best result of the security codes over wire-tap channel I is the linear code $(n, n-k)$ which can provide the maximum security rate $R < -\log_2(1-p)$ if this code is an optimum error detection code whose undetected error probability meets the upper bound 2^{-k}, where p is the error probability of Eve's channel [7]. However, a few small classes of linear codes have been proved to have an undetected error probability satisfying the upper bound 2^{-k}. The known optimum error detection codes included Hamming codes, double error-correcting BCH codes and Golay codes [28].

2.4.2 Unconditional Security Communication Model

In this section, we present our unconditional secure communication system by combining our building wiretap channel process and security codes together, which

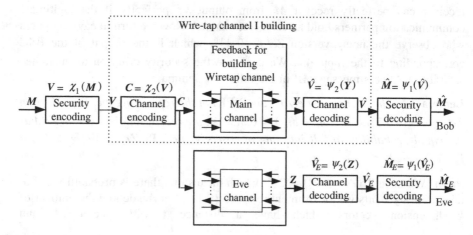

Fig. 2.6 Unconditional secure communication system model

is shown in Fig. 2.6. Alice wants to send k bits message $M^j = \{m_1^j, m_2^j, \cdots m_k^j\}$, $j = 1, 2, \cdots, 2^k$ to Bob. Firstly, Alice encodes the message such that

$$V = \chi_1(M), V \in \{0,1\}^{n_1} \tag{2.21}$$

where χ_1 is the security encoder function. Alice continues to encode V such that

$$C = \chi_2(M), C \in \{0,1\}^{n_2} \tag{2.22}$$

where χ_2 is the channel encoder function. Then Alice and Bob perform feedback process from Eq. (2.7) to Eq. (2.14). After several rounds of two-way communication between Alice and Bob (how many rounds need to be performed depends on the channel noisy level), Bob received the sequence Y which is the noisy version of sequence C. At the same time, the Eve can also observe the noisy sequence Z. Bob and Eve perform channel decoding as:

$$\begin{aligned} \hat{V} &= \psi_2(Y) \\ \hat{V}_E &= \psi_2(Z) \end{aligned} \tag{2.23}$$

where ψ_2 is the channel decoding function, which is an invertible function of channel encoding function χ_2. We have $Pr(Y \neq C) = p_1$, and $Pr(Z \neq C) = p_2$ where $p_2 > p_1$. By channel decoding, Bob can recover the sequence C as $\hat{V}$ perfectly and Eve can only get a noisy version of sequence C as $\hat{V}_E$, such that $Pr(\hat{V} \neq C) \to 0$ and $Pr(\hat{V}_E \neq C) \to \varepsilon > 0$. So the wiretap channel I have been built.

Then security decoding is performed as following:

$$\begin{aligned} \hat{M} &= \psi_1(\hat{V}) \\ \hat{M}_E &= \psi_1(\hat{V}_E) \end{aligned} \tag{2.24}$$

where ψ_1 is the security decoding function, which is an invertible function of security encoding function χ_1. By security decoding, Bob gets the message estimation $\hat{M}$ from $\hat{V}$ with error probability $Pr(\hat{M} \neq M) \to 0$ and Eve gets the message estimation $\hat{M}_E$ from V_E with error probability $Pr(\hat{M}_E \neq M) \to 0.5$ at the same time. Therefore, the security of Alice's message M is guaranteed.

2.4.3 Some Performance Results

We use parallel concatenated LDPC (PC-LDPC) codes in our performance. PC-LDPC code [25] is a kind of rate compatible codes which can easily adjust code rate to adapt to varying channel quality. The parity-check matrix H of the PC-LDPC codes has following form:

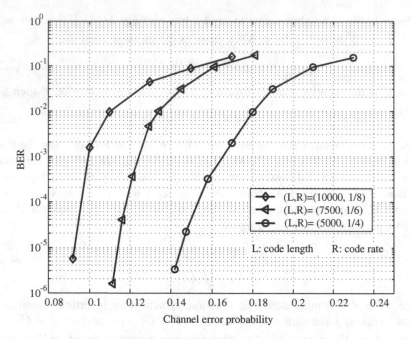

Fig. 2.7 The BER performances of LDPC codes

$$
H = \begin{bmatrix}
H_1^d & H_1^p & O & \cdots & O \\
H_2^d & O & H_2^p & \cdots & O \\
\vdots & \vdots & \vdots & \ddots & \vdots \\
H_s^d & O & O & \cdots & H_s^p
\end{bmatrix}
\tag{2.25}
$$

We employ seven random LDPC codes with rate $1/2$ as component codes. The overall code (mother code) length and rate are 10000 and $1/8$, respectively. The source block has 1250 bits. The component matrixes H_i^d and H_i^p in (2.25) are 1250 columns and 1250 rows with 3 weights for every column and row. By puncture the component matrixes H_i^p from the mother code matrix H, we can get the LDPC codes with length from 2500 to 10000 and the rate from $1/2$ to $1/8$.. The belief propagation iterative decoding algorithm is used and maximum number of iterations is 200. The performance of these codes is shown in Fig. 2.7. We launch the scenarios with two rounds (i.e., $t = 2$) of two-way communication. The bits error rate (BER) after LDPC codes decoding process are given in Table 1.4, where P_{ir} and P_{Eve} denote the BER of the intended receiver and the eavesdropper after decoding.

We take double error-correcting BCH codes as our security codes. Let the error probability of E_i be denoted as $Pr\left(e_i^j = 1\right) = \alpha_i$ and the error probability of Ee_i be denoted as $Pr\left(e_{ei}^j = 1\right) = \beta_i$. Two rounds (i.e., $t = 2$) of two-way communication are performed. The security code parameters are shown in Table 2.4. We present

Table 2.4 The performance properties with $t = 2$

Crossover probability	$\alpha_1 = \alpha_2 = 0.04$ $\beta_1 = \beta_2 = 0.02$	$\alpha_1 = \alpha_2 = 0.04$ $\beta_1 = \beta_2 = 0.04$	$\alpha_1 = \alpha_2 = 0.06$ $\beta_1 = \beta_2 = 0.04$	$\alpha_1 = \alpha_2 = 0.06$ $\beta_1 = \beta_2 = 0.06$	$\alpha_1 = \alpha_2 = 0.08$ $\beta_1 = \beta_2 = 0.04$	$\alpha_1 = \alpha_2 = 0.08$ $\beta_1 = \beta_2 = 0.08$
LDPC code	(5000, 1/4)	(5000, 1/4)	(5000, 1/6)	(7500, 1/6)	(10000, 1/8)	(10000, 1/8)
BER after interaction	$P(Y) = 0.0768$ $P(Z) = 0.11$	$P(Y) = 0.0768$ $P(Z) = 0.1418$	$P(Y) = 0.1128$ $P(Z) = 0.1723$	$P(Y) = 0.1128$ $P(Z) = 0.2002$	$P(Y) = 0.1472$ $P(Z) = 0.201$	$P(Y) = 0.1472$ $P(Z) = 0.251$
BER after CC decoding	$P_{ir} < 1.5 \times 10^{-5}$ $P_{Eve} = 0.0102$	$P_{ir} < 1.5 \times 10^{-5}$ $P_{Eve} = 0.073$	$P_{ir} < 1.16 \times 10^{-5}$ $P_{Eve} = 0.0125$	$P_{ir} < 1.16 \times 10^{-5}$ $P_{Eve} = 0.0172$	$P_{ir} = 1.9 \times 10^{-5}$ $P_{Eve} = 0.056$	$P_{ir} < 1.9 \times 10^{-5}$ $P_{Eve} = 0.195$
Cs	0.0819	0.3768	0.5434	0.6621	0.3110	0.7115
BCH code	(511, 493, 5)	(63, 51, 5)	(63, 51, 5)	(31, 21, 5)	(127, 113, 5)	(31, 31, 5)
BER after SC decoding	$P_B < 7.25 \times 10^{-3}$ $P_E = 0.4997$	$P_B < 5.35 \times 10^{-4}$ $P_E = 0.4999$	$P_B < 3.5 \times 10^{-4}$ $P_E = 0.4996$	$P_B < 6.2 \times 10^{-3}$ $P_E = 0.4999$	$P_B < 9.1 \times 10^{-4}$ $P_E = 0.4993$	$P_B < 5.2 \times 10^{-4}$ $P_E = 0.4995$
Rater of SC	0.0352	0.1905	0.1905	0.3226	0.1102	0.3226

the BER results after security decoding in Table 1.4, where P_B and P_E denote the BER of Bob and Eve after decoding, respectively. **SC** and **CC** is the logogram of security code and channel code. When the BER of Eve after **SC** decoding is less than 0.49, Eve is kept from receiving the information. From Table 1.4, we can conclude that the errors of Eve's received messages are higher than 0.49 and the errors of Bob's received messages are less than 10^{-3} even if the Eve has a same or better channel at the beginning.

Table 2.4 also shows the upper bound on the secret capacity C_s calculated according to lemma 2.1 in [4]. Although lower than the optimal secret capacity, the proposed security code is considered of practical use in realizing the unconditional security system. If we define the communication rate of a system as useful secret information bits via total transmission bits, our communication rate is higher than those of existing unconditional security schemes. In [15], it takes about 30 s to generate a 128 bits length secret key, which means that the total communication rate is about 7.2×10^{-7}. If our method is used to generate this length key, fourteen rounds (i.e., t = 14) of two-way communication need to be performed. $1/8$ code rate LDPC and 0.11 security code rate are used. If we estimate the overhead of communication as 25 %, the total communication rate is 2.45×10^{-5}. The other advantage of our scheme is that the multiple rounds two-way communication can be performed in parallel.

2.5 Summary

In this chapter, we have shown that the perfect security communication can be achieved by two steps. The first step is building wire tap channel by combining the feedback and LDPC codes. The second step is to extend the advantage of legitimate partners by the security codes. Instead of raising any unpractical assumption on the wiretapper's channels, the proposed approach in this chapter is applicable in any scenario with known noisy channels from the sender to the intended receiver and to the wiretapper, respectively. Thus, the proposed approach yields practical usage in many circumstances, such as on achieving confidentiality in a symmetric cryptographic system for key exchange, distribution, and message confidentiality.

Chapter 3
MIMO Based Enhancement for Wireless Communication Security

Using channel state information (CSI) as the secret key in multiple-inputmultiple-output (MIMO) links is another approach to guarantee low probability of interception (LPI) and realize physical layer (PHY) security techniques. However, the information-theoretic security is an average-information measure. The system can be designed and tuned for a specific level of security e.g., with very high probability a block is secure, but it may not be able to guarantee security with probability 1. The security in classical cryptography system is based on unproven assumptions regarding the hardness of certain computational tasks, therefore, systems are insecure if assumptions are wrong or if efficient attacks are developed. But the classical cryptography system can permit secure with probability 1 under some computing time and resource constraints.

Therefore, Innovative cross-layer security designs with both physical-layer security and upper-layer traditional security techniques are desirable for wireless networks. In this chapter, we propose a novel scheme with a cross-layer approach to enhance the security of wireless networks for wireless environments. We combine cryptographic techniques implemented in the higher layer with the physical layer security scheme using redundant antennas of MIMO systems to provide stronger security for wireless networks. By introducing a distort signal set instead of an orthogonal codeset for wireless networks based on space time block code (STBC) system [35–37], the transmitter randomly flip-flops between the distort signal set and the orthogonal code set for confusing the attacker. An upper-layer pseudorandom sequence will be employed to control the flip-flops process. In this approach the physical-layer can utilize upper-layer encryption techniques for security, while physical-layer security techniques can also assist the security design in the upper-layer.

H. Wen, *Physical Layer Approaches for Securing Wireless Communication Systems*, SpringerBriefs in Computer Science, DOI: 10.1007/978-1-4614-6510-2_3, © The Author(s) 2013

3.1 MIMO Cross-Layer Secure Communication Based on STBC

The STBC technique is a special form of physical layer diversity, which is a complex combination of coding theory, matrix algebra and signal processing. It is a technique, which operates on a block of input symbols producing a matrix and outputs whose columns and rows represent time and antennas, respectively. A key feature of STBC is the provision of full diversity with extremely low encoder/decoder complexity [35–37]. Therefore, STBC can be effectively used to exploit the advantage of MIMO systems. Recently, STBC techniques are suggested in the upcoming 802.11n standard proposals [36] by considering that STBC can help to achieve the full diversity and hence increase the data rates and throughput of an 802.11n system. In this section, we present a research for a cross-layer security scheme for STBC system. By introducing a distort signal set the sender randomly flip-flops between the distort signal set and the orthogonal code set for confusing the attacker. We illustrate our idea as following.

3.1.1 Overview of Alamouti Code

To explain our basic idea, we take the Alamouti STBC [37] as an example. Alamouti code is a special kind of STBC with two transmitters and one receiver as shown in Fig. 3.1. First, the transmitter picks two symbols from the constellation, where the signal constellation set consists of $M = 2^m$. If s_1 and s_2 are the selected symbols, the transmitter sends s_1 from antenna one and $-s_2^*$ from antenna two at time one. Then at time two, it transmits s_2 and s_1^* from antennas one and two, respectively. Therefore, the transmitted codeword is

$$s = \begin{pmatrix} s_1 & s_2 \\ -s_2^* & s_1^* \end{pmatrix} \tag{3.1}$$

where * denotes the conjugate transpose of the vector and the matrix. Let us assume that the path gains from transmit antennas one and two to the receive antenna are $h_1 = \alpha_1 e^{j\theta_1}$ and $h_2 = \alpha_1 e^{j\theta_2}$, respectively. Then the decoder receives signals r_1 and r_2 at time one and two, respectively, such that

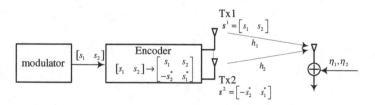

Fig. 3.1 Block diagram for Alamouti code

$$\begin{cases} r_1 = h_1 s_1 - h_2 s_2^* + \eta_1 \\ r_2 = h_1 s_2 + h_2 s_1^* + \eta_2 \end{cases} \tag{3.2}$$

where η_1, η_2 are Gaussian noise with zero-mean. If the receiver knows the channel path gains h_1 and h_2, the maximum-likelihood detection amounts to minimize the decision metric is

$$\left| r_1 - h_1 s_1 + h_2 s_2^* \right|^2 + \left| r_2 - h_1 s_2 + h_2 s_1^* \right|^2 \tag{3.3}$$

over all possible values of s_1 and s_2, where $| * |$ denotes the Euclidean norm of the vector. Substituting Eq. (3.2) into Eq. (3.3), the maximum likelihood decoding can be represented as

$$(\hat{s}_1, \hat{s}_2) = \min_{(\hat{s}_1,\hat{s}_2)\in S} \left\{ \left(|h_1|^2 + |h_2|^2 - 1 \right)\left(|\hat{s}_1|^2 + |\hat{s}_2|^2 \right) + d^2(\tilde{s}_1, \hat{s}_1) + d^2(\tilde{s}_2, \hat{s}_2) \right\} \tag{3.4}$$

where $(\hat{s}_1, \hat{s}_2)$ is the estimation value of the transmit signal pair (s_1, s_2); S is the set of all possible modulated symbol pairs $(\hat{s}_1, \hat{s}_2)$. $\tilde{s}_1$ and $\tilde{s}_2$ are two decision statistics given by

$$\begin{cases} \tilde{s}_1 = h_1^* r_1 + h_2 r_2^* \\ \tilde{s}_2 = h_1^* r_2 + h_2 r_1^* \end{cases} \tag{3.5}$$

For M-PSK modulation with equal energy constellations, Eq. (3.4) becomes:

$$\begin{aligned} \hat{s}_1 &= \arg\min_{\hat{s}_1 \in S}\left\{ d^2(\tilde{s}_1, \hat{s}_1) \right\} \\ \hat{s}_2 &= \arg\min_{\hat{s}_2 \in S}\left\{ d^2(\tilde{s}_2, \hat{s}_2) \right\} \end{aligned} \tag{3.6}$$

3.1.2 Distort Signal Set Design for Security Purpose

Instead of transmitting the signal in (3.1), we let the transmit signal set be

$$s = \begin{pmatrix} s_1 & s_2 \\ -s_3^* & s_4^* \end{pmatrix} \tag{3.7}$$

Let's consider identical channel fading coefficient $h_1 = h_2 = h_1 = h_2 = he^{j\theta}$ and the transmitters know the channel fading coefficients perfectly. When $s_3 = s_2$ and $s_4 = s_1$, Eq. (3.7) becomes Alamouti codeword in Eq. (3.1). So two decision statistics $\tilde{s}_1$ and $\tilde{s}_2$ in Eq. (3.5) become:

$$\begin{cases} \hat{s}_1 = \left(|h_1|^2 + |h_2|^2 \right) s_1 + h_1^* \eta_1 + h_2 \eta_2^* \\ \hat{s}_2 = \left(|h_1|^2 + |h_2|^2 \right) s_2 + h_1^* \eta_2 - h_2 \eta_1^* \end{cases} \qquad (3.8)$$

If we ignore the noise terms, the detection result will be $(\hat{s}_1, \hat{s}_2) = (s_1, s_2)$ under the maximum likelihood detection rule.

If $s_3 = s_1$ and $s_4 = s_2$, two decision statistics $\tilde{s}_1$ and $\tilde{s}_2$ become:

$$\begin{cases} \hat{s}_1 = \left(|h_1|^2 + |h_2|^2 \right) s_2 + h_1^* \eta_1 + h_2 \eta_2^* \\ \hat{s}_2 = \left(|h_1|^2 + |h_2|^2 \right) s_1 + h_1^* \eta_2 - h_2 \eta_1^* \end{cases} \qquad (3.9)$$

So the detection result will be $(\hat{s}_1, \hat{s}_2) = (s_1, s_2)$ under the maximum likelihood detection rule, while the actual signals sent by the transmitters are (s_1, s_2). This illustrates the ambiguity in the conventional signal detector when we send the distort signal set in some time slots. In our scheme an upper layer sequence set will be used to control that an Alamouti codeword or a distort signal set should be sent. Let the control sequence be $Q_{control} = (q_1, q_2, \cdots, q_n), q_i \in GF(2)$. Then

$$\begin{cases} if \quad q_i = 0, \quad \text{codeword in (2.1) is sent;} \\ if \quad q_i = 1, \quad \text{the distort signal is sent.} \end{cases} \qquad (3.10)$$

The control sequence $Q_{control}$ will be the secret key stream between the transmitter and the intended receiver. In general cases, if the transmit signal is the distort signal set in Eq. (2.7), two decision statistics $\tilde{s}_1$ and $\tilde{s}_2$ in (3.5) become:

$$\begin{cases} \tilde{s}_1 = |h_1|^2 s_1 - h_1^* h_2 s_3^* + h_2 h_1^* s_2^* + |h_2|^2 s_4 + h_1^* \eta_1 + h_2 \eta_2^* \\ \tilde{s}_2 = |h_1|^2 s_2 + h_1^* h_2 s_4^* - h_2 h_1^* s_1^* + |h_2|^2 s_3 + h_1^* \eta_2 + h_2 \eta_1^* \end{cases} \qquad (3.11)$$

Let the detection result be $\left(s_1^t, s_2^t \right)$ under the maximum likelihood detection rule, while the actual selected signals by the transmitters are s_1, s_2. We let

$$\begin{cases} \tilde{s}_1 = |h_1|^2 s_1 - h_1^* h_2 s_3^* + h_2 h_1^* s_2^* + |h_2|^2 s_4 + h_1^* \eta_1 + h_2 \eta_2^* = \left(|h_1|^2 + |h_2|^2 \right) s_1^t \\ \tilde{s}_2 = |h_1|^2 s_2 + h_1^* h_2 s_4^* - h_2 h_1^* s_1^* + |h_2|^2 s_3 + h_1^* \eta_2 + h_2 \eta_1^* = \left(|h_1|^2 + |h_2|^2 \right) s_2^t \end{cases}$$

$$\qquad (3.12)$$

By ignoring the noise, we have

$$\begin{cases} \tilde{s}_1 = |h_1|^2 s_1 - h_1^* h_2 s_3^* + h_2 h_1^* s_2^* + |h_2|^2 s_4 = \left(|h_1|^2 + |h_2|^2 \right) s_1^t \\ \tilde{s}_2 = |h_1|^2 s_2 + h_1^* h_2 s_4^* - h_2 h_1^* s_1^* + |h_2|^2 s_3 = \left(|h_1|^2 + |h_2|^2 \right) s_2^t \end{cases} \qquad (3.13)$$

Table 3.1 y_k corresponding the different choose of pair (c_i, c_j)

y_1	(c_1, c_1)	y_5	(c_5, c_5)	y_9	(c_9, c_9)	y_{13}	(c_{13}, c_{13})
y_2	(c_2, c_2)	y_6	(c_6, c_6)	y_{10}	(c_{10}, c_{10})	y_{14}	(c_{14}, c_{14})
y_3	(c_3, c_3)	y_7	(c_7, c_7)	y_{11}	(c_{11}, c_{11})	y_{15}	(c_{15}, c_{15})
y_4	(c_4, c_4)	y_8	(c_8, c_8)	y_{12}	(c_{12}, c_{12})	y_{16}	(c_{16}, c_{16})

To maximize the ambiguity, we should maximize the probability of detection error by designing the distort signal pair (s_3, s_4) carefully. Therefore, the pair (s_1^t, s_2^t) should satisfy

$$\hat{s}_1^t = \arg\max_{\hat{s}_1^t \in S}\left\{d^2\left(s_1, \hat{s}_1^t\right)\right\}$$
$$\hat{s}_2^t = \arg\max_{\hat{s}_2^t \in S}\left\{d^2\left(s_2, \hat{s}_2^t\right)\right\} \tag{3.14}$$

To illustrate how to choose the distort signal pair (s_3, s_4). Let us consider the following example. We suppose that the signaling is through QPSK with constellations $C = (c_1, c_2, c_3, c_4)$ and the received signal is noise free with the channel fading coefficient $h_1 = 1$ and $h_2 = e^{j\frac{\pi}{4}}$ i.e., $y_k = h_1 c_i + h_2 c_j$, $1 \le i, j \le 4$, where y_k can take 16 distinct patterns corresponding to the different choose of pair (c_i, c_j) and is shown in Table 3.1 and Fig. 3.2. When the received signal contains additive noise, the received signal takes the form $r = y_k + \eta$, where η is the complex Gaussian noise. The signal pair (c_i, c_j) can be detected from r. Under the complex Gaussian noise, the maximum likelihood detector is equivalent to the minimum distance detector. To maximize the ambiguity, we should let pairs (s_1^t, s_2^t) and (s_1, s_2) take the signal pair (c_i, c_j) to maximize the Euclidean distance between (s_1^t, s_2^t) and (s_1, s_2). For example, if the transmitter picks up the signal constellation pair (c_1, c_4) at the first time slot, to maximize the probability of attackers' detection errors, we let the pair (s_1^t, s_2^t) take (c_3, c_2). So we can solve the distort signal pair (c_3, c_4) as $\left(\sqrt{2} + (1 + \sqrt{2})j, -1 + \sqrt{2} - \sqrt{2}j\right)$ from (3.13).

3.1.3 The Action of the Legitimate Receiver

The legitimate receiver knows the control sequence. Therefore, he will detect the signals via Eq. (3.2) to (3.6) when the sender transmits the signals according to Eq. (3.1). If the sender sends the distort signals, he will detect the signals as followings.

The legitimate decoders receive signals r_1 and r_2 at times one and two, respectively, such that

Fig. 3.2 The distinct patterns combining signal constellation with the channel fading coefficient $h_1 = 1$ and $h_2 = e^{j\frac{\pi}{4}}$

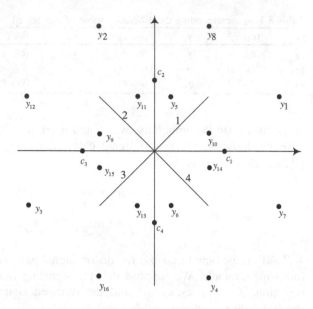

$$\begin{cases} r_1 = h_1 s_1 - h_2 s_3^* + \eta_1 \\ r_2 = h_1 s_2 + h_2 s_4^* + \eta_2 \end{cases} \tag{3.15}$$

Then the decoders calculate two decision statistics $\tilde{s}_1$ and $\tilde{s}_2$ as

$$\begin{cases} \tilde{s}_1 = h_1^* r_1 + h_2 r_2^* \\ \tilde{s}_2 = h_1^* r_2 + h_2 r_1^* \end{cases} \tag{3.16}$$

Let the pair $(\hat{s}_1^t, \hat{s}_2^t)$ denote the estimation of the pair (s_1^t, s_2^t). For equal energy constellations modulation, we detect the estimation pair $(\hat{s}_1^t, \hat{s}_2^t)$ as

$$\begin{aligned} \hat{s}_1^t &= \arg\min_{\hat{s}_1^t \in S}\{d^2(\tilde{s}_1, \hat{s}_1^t)\} \\ \hat{s}_2^t &= \arg\min_{\hat{s}_2^t \in S}\{d^2(\tilde{s}_2, \hat{s}_2^t)\} \end{aligned} \tag{3.17}$$

Finally, we can get the detection results of transmit signal pair (s_1, s_2) as

$$\begin{aligned} s_1^t &= \arg\max_{\hat{s}_1 \in S}\{d^2(\hat{s}_1^t, \hat{s}_1)\} \\ s_2^t &= \arg\max_{\hat{s}_2 \in S}\{d^2(\hat{s}_2^t, \hat{s}_2)\} \end{aligned} \tag{3.18}$$

For an attacker, he does not know the control sequence. So he does not know when the Alamouti code pair are sent and when the distort signal pair are sent. For a pair of codewords $(\hat{s}_1, \hat{s}_2)$ and $(\hat{s}_1^t, \hat{s}_2^t)$, let us define the Euclidean distance between the codewords, denoted by d_E, as

$$d_E = \sqrt{\left|\hat{s}_1 - \hat{s}_1^t\right|^2 + \left|\hat{s}_2 - \hat{s}_2^t\right|^2} \tag{3.19}$$

An attacker still detects the signals via Eq. (3.2) to (3.6) when the distort signals are sent. d_E will become additional noise.

3.2 Security Performance Analysis

3.2.1 Attackers Action with Multiple Receive Antennas

It is possible that the attackers use more than one antenna to receive the signal. In this section, we analyze the system security in this scenario. To explain eavesdroppers action with multiple receive antennas, we take the Alamouti scheme with two transmit and two receive antennas as an example as shown in Fig. 3.3. Let us denote by r_1^{Ej} and r_2^{Ej} the eavesdroppers' received signals at the jth receive antenna at time slot 1 and 2, respectively.

$$\begin{aligned} r_1^{Ej} &= h_{j,1}s_1 + h_{j,2}s_2 + \eta_1^j \\ r_2^{Ej} &= h_{j,1}s_2^* + h_{j,2}s_1^* + \eta_2^j \end{aligned}, \quad j = 1, 2 \tag{3.20}$$

where $h_{j,i}, j, i = 1, 2$ is the fading coefficient for the path from transmit antenna i to receive antenna j, and η_1^j, η_2^j are the noise signals for receive antenna j at time slot 1 and 2, respectively. The receiver constructs two decision statistics, denoted by $\tilde{s}_1^{Ej}$ and $\tilde{s}_2^{Ej}$ are given by

Fig. 3.3 Attacker actions with additional antenna

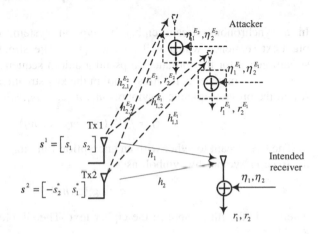

$$\tilde{s}_1^{Ej} = \sum_{j=1}^{2} \left[h_{j,1}^* r_1^{Ej} + h_{j,2} \left(r_2^{Ej} \right)^* \right] = \sum_{i=1}^{2} \sum_{j=1}^{2} \left| h_{j,i} \right| s_1 + \sum_{j=1}^{2} \left[h_{j,1}^* \eta_1 * j + h_{j,2} \left(\eta_2^j \right)^* \right]$$

$$\tilde{s}_2^{Ej} = \sum_{j=1}^{2} \left[h_{j,2}^* r_1^{Ej} + h_{j,1} \left(r_2^{Ej} \right)^* \right] = \sum_{i=1}^{2} \sum_{j=1}^{2} \left| h_{j,i} \right| s_2 + \sum_{j=1}^{2} \left[h_{j,2}^* \eta_1 * j + h_{j,1} \left(\eta_2^j \right)^* \right]$$

$$(3.21)$$

Let the pair (s_1, s_2) be the signal pair without distort signals. The maximum likelihood decoding for the two independent signals and can be represented as

$$\hat{s}_1 = \arg \min_{\hat{s}_1 \in S} \left\{ \left[\sum_{j=1}^{2} \left(\left| h_{j,1} \right|^2 \right) + \left| h_{j,21} \right|^2 - 1 \right] |\hat{s}_1|^2 + d^2(\tilde{s}_1, \hat{s}_1) \right\}$$

$$\hat{s}_2 = \arg \min_{\hat{s}_2 \in S} \left\{ \left[\sum_{j=1}^{2} \left(\left| h_{j,1} \right|^2 \right) + \left| h_{j,21} \right|^2 - 1 \right] |\hat{s}_2|^2 + d^2(\tilde{s}_2, \hat{s}_2) \right\}$$

$$(3.22)$$

Let the pair $\left(s_1^t, s_2^t \right)$ be true sent signal pair which include the distort signals, and the pair $\left(\hat{s}_1^t, \hat{s}_2^t \right)$ denote the estimation of the pair $\left(s_1^t, s_2^t \right)$. For an attacker, he does not know the control sequence. He does not know when the Alamouti code pair are sent and when the distort signal pair are sent. Therefore, he can not recover the pair $\left(s_1^t, s_2^t \right)$. Similar to the scenario where the attacker with one receive antenna, the Euclidean distance between the pair $\left(\hat{s}_1^t, \hat{s}_2^t \right)$ and the pair $\left(\hat{s}_1^t, \hat{s}_2^t \right)$ becomes the additional noise. The additional antenna can not improve the attack receive performance.

3.2.2 Comparing the Proposed New Models with the Traditional Stream Cipher Encryption System

In a synchronous stream cipher encryption system, let $m = m_1, m_2 \ldots$ be a plaintext sequence which will be encrypted. The stream cipher contains a key stream generator that produces a pseudorandom sequence, called the key stream, $z = z_1, z_2 \ldots$ In general, the ith symbol in the key stream, z_i, is a function of the key K and the previous plaintext symbols $m_1, m_2, \ldots, m_i$, which can be represented by

$$z_i = f_i[K, (m_1, m_2, \ldots, m_i)] \qquad (3.23)$$

The key stream together with an encryption function, g_i, are used to encrypt the message m symbol by symbol, as

$$c_i = g_i(z_i, m_i) \qquad (3.24)$$

where c_i is the ith symbol in the cipher text. The simplest case is that

$$z_i = f_i[K] \tag{3.25a}$$

and

$$c_i = z_i \oplus m_i \tag{3.25b}$$

Usually, we hope that the functions f_i and g_i are nonlinear functions. In practical application of stream cipher encryption systems, we use different kind of nonlinear functions to imitate the random properties. In our new model, let s_1' and s_2' denote the first four terms in (3.11), we have

$$s_j' = fc2_j^i[Q_{control}, (s_1, s_2, \ldots, s_i)] \tag{3.26}$$

The decision statistics in (3.22) can model as:

$$\tilde{s}_j = gc2_j^i \left[s_j', (s_1', s_2', \ldots, s_i') \right] \tag{3.27}$$

In Eqs. (3.26) and (3.27), the functions $fc2_j^i$, $gc2_j^i$ are determined by fading channel coefficients and noises which are random functions with high nonlinear properties. These random and nonlinear properties can be broken by channel estimation or searching the controlling sequences. But the performance properties BER have to be paid for the attacker. The secret capacity can be obtained. We also compare the attacking complexity of our scheme with those of traditional stream cipher encryption systems.

Here we only consider the brute force attacking situation and the known plan text attack. In the proposed scheme, $Q_{control}$ is pre-shared key stream between the transmitter an intended receiver and the attack tries to find the secret key of $Q_{control}$ from the $\hat{S}$ which is the attacker's estimate of S in (3.1). We have

$$\hat{S} = S + E \tag{3.28}$$

where E is the error matrix. If $E = O$ we have $\hat{S} = S$. One can get $Q_{control}$ easily and the methods of finding secret key from the key stream $Q_{control}$ have lots of research results [33, 34]. The attacker can also use the brute force search method to find the secret key of $Q_{control}$ by testing the 2^k different secret keys if we assume that the secret key is k bits, which is the brute force attacking complexity of traditional stream cipher encryption system. If the probability of $\Pr(\hat{S} \neq S)$ is high, the attacker has two brute force methods to find the secret keys. One method is that the attacker exhausts the entire secret key to construct the $Q_{control}$. Then he uses the $Q_{control}$ to do receiving procedure. If the secret key is k bits, the attacker has to do 2^k times receiving procedure from (3.15) to (3.18). Another method is that the attacker guesses the error matrix E and tries to find the control sequences. If the elements of E is B_n, the error rate is R_e and values that the elements of E can take are M, the exhaustive search number will be

$$\sum_{i=0}^{B_n \cdot R_e} 2^k \cdot \binom{B_n}{i} \cdot (M-1)^i \tag{3.29}$$

3.3 The Simulation Performance of the Proposed Method

3.3.1 The Performance Properties

In this section, we simulated the effectiveness of the proposed transmission scheme by evaluating the bit error rate (BER) of the intended receiver and the attacker under 802.11n system [36]. The channel is assumed to be able to block Rayleigh fading, i.e., it is constant during the transmission of one packet, but randomly changes between packets.

BER performance of the intended receiver and the attacker are measured under Alamouti code scheme with two transmit antennas and one receive antenna. In the following simulation, it is assumed that the attackers know the main channel fading coefficients h_1 and h_2 between legitimate patterns. BPSK, QPSK and 8-PSK modulation are employed, respectively. The WG stream ciphers [33] are employed to generate the control sequences.

Firstly, it is assumed that the transmitter and receiver know the perfect channel state information. The simulation result in Fig. 3.4 shows that the attacker can only receive the signal with almost half errors when the intended receiver can receive the signal with errors less than a BER of 10^{-4}. Further, the situation that the attackers are located very close to the intended receiver is considered. The performances are evaluated when the distance between the attackers and the legitimate users is 1/2 and 1/4 times wavelength (WL). Figure 3.5 illustrates the results of BER performance as a function of Signal-to-Noise Ratio (SNR). The BER performance results of the attacker with two Receive Antennas (RAs) are shown in Fig. 3.6, from which we can conclude that the additional antenna can not improve the attack receive performance.

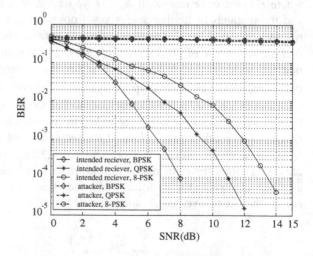

Fig. 3.4 BER performance of cross-layer secure communication architecture based on STBC

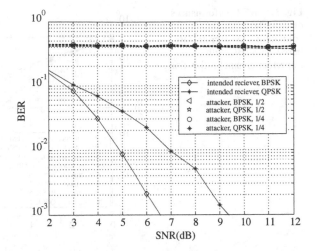

Fig. 3.5 BER performances while the distance between the attackers and the legitimate users is very close

Secondly, we consider imperfect channel state information at both the receiver and transmitter. We measure the BER performance of the intended receiver and the attacker with the channel estimation method used in the simulation [38]. Then the Channel State Information (CSI) is estimated by inserting pilot sequences in the transmitted signals. It is assumed that the channel is constant over the duration of a frame and independent between frames. The frame length is 130 symbols and the pilot sequence inserted in each packet has a length of 12 symbols. Performances are illustrated in Fig. 3.7, in which 'IC' denotes ideal channel information and 'NC' denotes non-ideal channel information. The simulation result shows that due to imperfect channel estimation, the performance of both the receiver and transmitter is degraded, but the intended receiver still can receive signal with low BER.

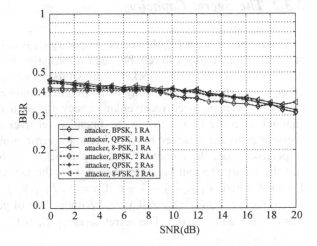

Fig. 3.6 BER performances of the attacker with two receive antennas

Fig. 3.7 BER performances
of cross-layer secure
communication architecture
based on STBC with
imperfect channel estimation

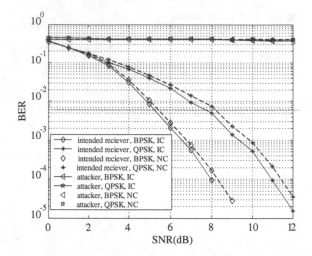

The secret keys of the WG stream ciphers [33] that are used as the noise sequences and control sequences are 35 bits which is not long enough for traditional cryptographic system, but our system is different. The attacker has to find secret key from the noisy received signals, which is more difficult than traditional attacking to ciphers which assumed that the key stream is error free. From the simulation results, we can learn that the attacker can only receive the signal with almost half errors when the intended receiver can receive the signal with low errors under the simulation condition that the attackers know the main channel fading coefficients between legitimate patterns, which means that if an attacker is located very close to the receiver, our methods are still secure. Therefore, our approach can guarantee security with probability 1.

3.3.2 The Secret Capacity

In Csiszar and Korner [3], the secrecy capacity C_S is defined as the maximum rate at which a transmitter can reliably send information to an intended receiver such that the rate at which the attacker obtains this information is arbitrarily small. In other words, the secrecy capacity is the maximal number of bits that a transmitter can send to an intended receiver in secrecy for each use of the channel. If the channel from the transmitter to the intended receiver and the channel from the transmitter to the attacker have different bit error probabilities (BER) ε and δ, respectively, i.e., the common input to channel is the binary random variable X, and the binary random variables received by the legitimate and the attacker are Y and Z where $P_{Y|X}(y|x) = 1 - \varepsilon$ if $x = y$, $P_{Y|X}(y|x) = \varepsilon$ if $x \neq y$, $P_{Z|X}(z|x) = 1 - \delta$ if $x = z$, $P_{Z|X}(z|x) = \delta$ if $x \neq z$. Without loss of generality, we may assume that $\varepsilon \leq 0.5$ and $\delta \leq 0.5$. The secret capacity C_s is [7]:

Fig. 3.8 The secret channel capacity from the BER performance results based on STBC scheme

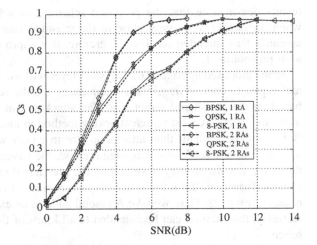

$$C_s = \begin{cases} h(\delta) - h(\varepsilon), & if \quad \delta > \varepsilon \\ 0, & otherwise \end{cases} \tag{3.30}$$

where h denotes the binary entropy function defined by

$$h(p) = -p \log_2 p - (1-p) \log_2 (1-p) \tag{3.31}$$

The secrecy capacity of the multi-antenna system is introduced in [32], which depends on the existence of the enhanced channel. This characterization is directly built on the optimal transmission strategy in the multi-antenna system.

Our schemes are not built on this approach. Therefore, we still use the simple result in (3.30) to evaluate the secret capacity of the new schemes.

Based on the BER results in Figs. 3.4 and 3.6, the secret capacity is calculated by (3.30) and shown in Fig. 3.8. The solid lines denote the secret capacity between the intended receiver and the attack with one receive antenna while the dash lines indicate the secret capacity between the intended receiver and the attack with two receive antennas. It is assumed that the transmitter and the intended receiver can achieve the normal communication when the BER performance of the intended receiver is less than 10^{-2}. Therefore, our new method can achieve sufficiently good secret capacity within the corresponding SNR ranges.

3.4 Summary

In this chapter, a MIMO-based secure communication scheme is presented. In this scheme, a random flip-flops process controlled by upper-layer pseudorandom sequence is employed to confuse the attacker at the physical layer. Therefore, this approach is a cross-layer security design method. The extensive performance results demonstrate that the proposed scheme can achieve secure performance when the attackers are very close to the intended user and have more receiving antennas than intended receiver. Therefore, this approach can guarantee security with probability 1.

In the proposed approach, the physical-layer will use the upper-layer encryption techniques to control transmission sequences. So the security of this approach is based on the pre-shared secret key. The attackers can also use certain attacking methods such as exhaust search attacking method to find the pre-shared secret key. But it is more difficult than attacks on stream cipher which assume that the key stream is error free. In our proposed method, the eavesdroppers can only receive the noisy signals, so they have to find the pre-shared secret key in the noisy key stream. Therefore, the introduced scheme provides stronger security for communication two parts. Here we take Alamouti code as an example to explain the idea. Actually, this method can be extended to all kinds of the space–time block codes schemes.

Chapter 4
Physical Layer Assisted Authentication for Wireless Sensor Networks

Real-time wireless broadcast is envisioned to take an important role in Wireless Sensor Networks (WSNs). For example, routing tree construction, network query, software updates, time synchronization, and network management all rely on broadcast. However, due to the nature of wireless communication in sensor networks, attackers can easily inject malicious data or alter the content of legitimate messages during multihop forwarding. Authentication of the broadcast messages is an effective approach to countermeasure most of the possible attacks, by which the intended receivers can make sure that the received data is originated from the expected source.

With physical layer authentication, a lightweight and fast wireless broadcast authentication scheme can be developed for, in which a channel response is extracted and used to estimate the channel information before the signals are received. In this chapter, we explore the potential possibility of using physical-layer channel responses as authenticators between each communication pair, and propose two novel message authentication schemes are called as Physical-layer Assisted Authentication (PAA) for Vehicular Ad hoc Networks (VANETs) and Ad Hoc Wireless Sensor Networks (AWSN), in order to cope with the stringent requirements on message authentication delay and communication overhead.

4.1 System Model

We consider the scenario illustrated in Fig. 4.1 by taking the conventional terminology widely used in the security community for three different parties in a secure multicast scenario: a sender, multiple receivers, and Eavesdroppers (Eves). These three entities could also be taken as a wireless transmitter and a number of receivers that are potentially located in spatially separated positions. In this book, the sender serves as the transmitter that initiates communication with the receivers, while Eve is an active opponent who injects undesirable signals into the medium in the hopes of spoofing the sender. Our security objective is to provide authentication between the

H. Wen, *Physical Layer Approaches for Securing Wireless Communication Systems*, SpringerBriefs in Computer Science, DOI: 10.1007/978-1-4614-6510-2_4, © The Author(s) 2013

sender and the legal receivers, in which the legal receivers have to differentiate a signal launched by the sender from that by Eve. To achieve this, each information-carrying transmission is accompanied with the channel response at the sender, which serves as an authenticator signal for the receivers to verify the legitimacy of the transmission. With the authenticator signal, each receiver authenticates the transmission by comparing the authenticator signal with the estimated channel response. If the two channel estimates are "close" to each other, the receiver will conclude that the source of the message is the same as the source of the previously sent message. If the channel estimates are not similar, then the receiver will reject the message.

Suppose the source sends a signal to the receivers with the frame structure shown in Fig. 4.2, where we consider a cyclic prefix (CP) Orthogonal Frequency-Division Multiplexing (OFDM) system with M subcarriers and total transmit bandwidth B. The duration of one OFDM symbol is T. A packet consists of frames N_x, which consist of N_d OFDM data symbols and one pilot in each subcarrier.

The corresponding system model is then given by

$$Y_k(n, m) = H_k(n, m)X_k(n, m) + Z_k(n, m) \tag{4.1}$$

where $k \in \{1, 2, \cdots, N_x\}$, $n \in \{1, 2, \cdots, N_d\}$ and $m \in \{1, 2, \cdots, M\}$ are the frame index, the symbol index and subcarrier index, respectively; furthermore, $Y_k(n, m)$ and $X_k(n, m)$ are the receive and transmit symbols, respectively, $H_k(n, m)$ denotes the channel coefficients, and $Z_k(n, m)$ is additive white Gaussian noise with variance σ_Z^2. We consider time-varying Rayleigh fading channel. The channel time–frequency domain coefficient $H_k(n, m)$ are related to the delay-Doppler spreading function $S_k(n, m)$ of the channel via a 2-D Fourier transform [49],

$$H_k(n, m) = \frac{1}{\sqrt{MN_d}} \sum_{\tau=0}^{M_\tau - 1} \sum_{l=-\frac{M_v}{2}}^{\frac{M_v}{2}} S_k(n, m) e^{-j2\pi(\frac{m\tau}{M} - \frac{nl}{N_d})} \tag{4.2}$$

where τ, l denote discrete delay and discrete Doppler, respectively. M_τ dentotes the channel's maximum delay spread and M_v characterizes the maximum Doppler spread. Since in practice $M_\tau \ll M$ and $M_v \ll N_d$, $H_k(n, m)$ is a 2-D lowpass function.

Fig. 4.1 Scenario with sender, receivers and Eves

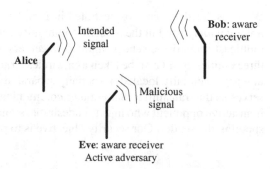

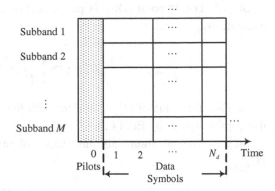

Fig. 4.2 Frame structure of the transmission from the sender to the receivers

Subband 1

Subband 2

⋮

Subband M

0   1   2 $\cdots$ N_d Time

Pilots ⟵———— Data ————⟶
                     Symbols

4.2 Physical Channel Identification

4.2.1 Physical Layer Channel Response Extraction

To recover the unknown channel coefficients from the pilot, linear and Iterative Least-Squares (ILS) adaptive [54], one-dimensional (1-D), and two-dimensional (2-D) Minimum Mean-Square Error (MMSE) filtering algorithms have been proposed [52]. Here we introduce ILS channel estimation method as following. For simplicity, we write $Y_k(n, m)$, $X_k(n, m)$, $H_k(n, m)$, and Z_k as Y_k, X_k, H_k, and Z_k, respectively.

Step 1: It is straightforward to obtain a LS channel estimate at the pilot positions according to

$$\hat{H}_{LS\_k}^1 = \frac{Y_k}{X_k} = H_k + \frac{Z_k}{X_k} \tag{4.3}$$

Then the following iterative algorithm is applied to the channel estimate $\hat{H}_{LS\_k}^1$.

Step 2: Let M be the number of symbols in a pilot. The transform domain representation of frequency domain channel estimate is expressed as:

$$\hat{S}_k^1 = \sum_{l=1}^{M} \hat{H}_{LS\_k}^1 e^{\frac{-j2\pi lp}{M}} \tag{4.4}$$

Step 3: Lowpass filter is performed on $\hat{S}_k$ simply by setting the samples in the "high frequency" region to zero, that is:

$$\hat{S}_k^1 = \begin{cases} \hat{S}_k^1, & 0 \leq p \leq p_c - 1 \\ 0 & p_c \leq p \leq N - p_c - 1 \\ \hat{S}_k^1, & N - p_c \leq p \leq N - 1 \end{cases} \tag{4.5}$$

where p_c is the "cutoff frequency" of the filter in the transform domain.

Step 4: The M-point IDFT is performed on $\hat{S}_k^1$ to gain the frequency channel estimate $\hat{H}_{LS\_k}^2$:

$$\hat{H}_{LS\_k}^2 = \sum_{p=1}^{M} \hat{S}_k^1 e^{\frac{j2\pi pm}{M}} \tag{4.6}$$

Replace the value of $\hat{H}_{LS\_k}^2$ in pilot with the initial LS channel estimate value using pilot signals by Eq. (4.3).

Step 5: The maximum absolute value of the difference between $\hat{H}_{LS\_k}^1$ and $\hat{H}_{LS\_k}^2$ is:

$$\delta = \max \left| \hat{H}_{LS\_k}^2 - \hat{H}_{LS\_k}^1 \right| \tag{4.7}$$

If δ falls below a predetermined threshold, the iteration should be terminated and the final decision is made; otherwise, take $\hat{H}_{LS\_k}^2$ as the new frequency channel estimate and go back to repeat step 2–5.

After step 1 to step 5 we get channel response H_k from the pilot.

4.2.2 Physical Layer Authentication Based on LRT

In this section, we briefly review some facts on the Physical Layer Authentication (PLA) scheme introduced by Xiao et al. [45–48]. The receivers use the pilots for channel estimation, which can come up with the test vectors as follows

$$\underline{\hat{H}}_k(n) = \left[\hat{H}_k(n, 1), \hat{H}_k(n, 2), \cdots, \hat{H}_k(n, M) \right]^T \tag{4.8}$$

where k is the frame index, M is the number of subcarriers and n is OFDM symbol index. For convenience, the index n is omitted. Then Eq. (4.3) can be rewritten as:

$$\underline{\hat{H}}_k = \left[\hat{H}_k(1), \hat{H}_k(2), \cdots, \hat{H}_k(M) \right]^T \tag{4.9}$$

The receiver uses channel estimates in two consecutive frames, $\underline{\hat{H}}_{k-1}$ and $\underline{\hat{H}}_k$, to determine whether they are from the same transmitter (sender) or not. The receiver is assumed to obtain the sender-receiver channel gain in the previous frame, $\underline{\hat{H}}_{k-1}$, and can compare it with the current channel gain, $\underline{\hat{H}}_k$, to determine whether or not the current received frame is from a common transmitter that has launched the previous one. $\underline{\hat{H}}_{k-1}$ and $\underline{\hat{H}}_k$ can be estimated by ILS and MMSE channel estimation methods. In the null hypothesis, H_0, the claimant is the original send. Otherwise, in the alternative hypothesis, H_1, the claimant terminal is someone else. The notation $\sim$ is used to denote accurate values without measurement error, and thus we have:

$$H_0 : \quad \tilde{\underline{H}}_k \rightarrow \tilde{\underline{H}}_{k-1}$$
$$H_1 : \quad \tilde{\underline{H}}_k \mapsto \tilde{\underline{H}}_{k-1} \tag{4.10}$$

We normalize the likelihood ratio test (LRT) statistic as:

$$\Lambda_0 = \frac{K_{co}\left\|\hat{\underline{H}}_k(i) - \hat{\underline{H}}_{k-1}(i)e^{j\varphi}\right\|^2}{\left\|\hat{\underline{H}}_{k-1}(i)\right\|^2} \quad \begin{matrix} > H_1 \\ < H_0 \end{matrix} \quad \eta \tag{4.11}$$

where $\hat{\underline{H}}_k$ represents the channel response with measurement errors; K_{co} is the normalization factor and $\eta \in [0,1]$.

As mentioned in Ref. [45], the authentication scheme depends on the richly scattered multipath environment. Because the received signal rapidly decorrelates over a distance of roughly half a wavelengths and the spatial separation of one to two wavelengths is sufficient for assuming independent fading paths, the time interval of two continuous authentication procedures should be less than the channel's coherence time τ.

4.2.3 Physical Layer Authentication Based on SPRT

We developed the PAA scheme by the sequential probability ratio test (SPRT), which was initially introduced by Abraham Wald [55] as a hypothesis test for sequential analysis. The SPRT considers sequential sample units as a statistical test. LRT only compares the estimation in the k-th frame ($\hat{\underline{H}}_k$) with that in the $(k-1)$-th frame ($\hat{\underline{H}}_{k-1}$). A sequential probability ratio test can compare ($\hat{\underline{H}}_k$) with all past records ($\hat{\underline{H}}_i$) where $i < k$ in some way which may yield a better detection rate. So it is possible that SPRT is better than the simple LRT. Let ith statistical test the log-likelihood ratio as:

$$\Lambda_i = \frac{K_{co}\left\|\hat{\underline{H}}_{k-i+1}(i) - \hat{\underline{H}}_{k-i}(i)e^{j\varphi}\right\|^2}{\left\|\hat{\underline{H}}_{k-i}(i)\right\|^2}, \quad i = 1, 2, \cdots, S \tag{4.12}$$

where $S \geq 1$. Then we calculate the cumulative sum of the log-likelihood ratio Λ as:

$$\Lambda = K_{co\_s}\sum_{i=1}^{S}\Lambda_i \quad \begin{matrix} > H_1 \\ < H_0 \end{matrix} \quad \delta \tag{4.13}$$

where K_{co_s} is the normalization factor to let the threshold $\delta \in [0,1]$. Because the channel response may change with time due to change in the environment, it is necessary to guarantee the time intervals of the continuous authentication procedure less than the channel's coherence time τ. The number of the statistical test cumulative sum S in Eqs. (4.12) and (4.13) has to be set as $t_k - t_{k-s} < \tau$, i.e. S time intervals is less than the channel's coherence time τ.

4.3 Physical Layer Assisted Authentication Based on PKI

This section will present the Physical layer Assisted Authentication (PAA) security scheme under PKI, which aims to achieve high efficiency in terms of packet overhead incurred in the message authentication scheme and computation latency.

4.3.1 Physical Layer Assisted Authentication Principle

Let all the entities be installed with a list of anonymous public/private key pairs $\langle PK_i, SK_i \rangle$, where the corresponding anonymous certificates are $Cert_i$ with pseudo identities $PVID_i$ as its certificate identities [59]. For the purpose of traceability, vehicle registration authority keeps records of those anonymous certificates and their corresponding real identities. Each pair of keys has a short life time, e.g., a few minutes. A sender sends the first massage to the receivers. A signature is produced for the first message with the conventional public key signature technique. Then the receiver estimates the channel response $\hat{H}_1$ by packet pilot which was originally designed to serve as the channel estimation and saves it. For the following messages, the receiver estimates the channel response $\hat{H}_i$ by packet pilot and compare with $\hat{H}_{i-1}$ according to Eq. (4.11) or (4.13). If the message verifies successfully, the receiver saves $\hat{H}_i$ instead of $\hat{H}_{i-1}$ for the next packet authentication. If the message does not verify successfully, the receiver drops the message and picks up another public/private key pairs and initiates new PAA scheme again.

Let the routine messages sent by a vehicle be denoted as $X_1, X_2, \cdots, X_N$, and $X_i(i = 1, 2, \cdots, N)$ is ith transmitting packet. The proposed security scheme is shown in Fig. 3.3. For an arbitrary sender V, before it sends the first message, it signs the hashed message with its private key SK_V and includes the certification authority's certificate $Cert_V$ as follows:

$$P_1 = \langle PVID, X_1, Sig_{SK_V}[X_1|T_1], \ Cert_V \rangle \tag{4.14}$$

where $PVID$ is the pseudo ID of the sender V, which is kept in accordance with the ID that is being used in the current public key certificate $Cert_V$ that is the currently used anonymous public key certificate; j is the concatenation operator, and T_1 is the time when the sender sends the packet, which is used to defeat replay attack. When the sender sends the following messages, he directly picks up the packet X_1 without any authentication tag.

The receivers of the first message have to extract the public key of V using the certificate, and then verify V's signature using its certified public key. The first message will go through a complete authentication process via the conventional PKI mechanism. With the first message fully authenticated, the receivers perform channel estimation, extract and save the channel responses with the sender $\hat{H}_1$. For each of the subsequent messages, the receivers perform channel estimation $\hat{H}_i$ according to the currently received message, and compare it with the channel response $\hat{H}_{i-1}, \hat{H}_{i-2}, \cdots, \hat{H}_{i-S}$ of the previously received S frame messages, where the physical layer authentication is performed according to Eq. (4.13). The physical layer authentication will be performed according to Eq. (3.11) if S equals to 1. In case the verification fails, the packet is dropped. Otherwise, the receiver updates the entry $(packet\#, \hat{H}_i, lifetime)$ in its local cache table corresponding to the sender V, which maintains the packet index, channel response, and lifetime which serves as a timer controlling how long the entry is active. If the timer hits 0, another freshness process in Eq. (3.14) has to be done again. As mentioned earlier, when the time interval of two continuous authentication procedures is larger than the channel's coherence time, the received signals decorrelates each other. So the receiver needs to check whether the time interval $t_i - t_{i-1}$ between two slots is larger than the channel's coherence time τ after the successful verification successfully, if the test statistic Eq. (4.11) is used. If we use Eq. (4.13) as the test statistic, the receiver needs to check whether the S larger than 1 after the successful verification successfully. When $t_i - t_{i-1} < \tau$ or $S \geq 1$ hold, the authentication procedure continues. Otherwise, the sender has to refresh the process in Eq. (4.14) and initiates a new PAA scheme process again. The proposed PAA authentication scheme is shown in Fig. 4.3.

4.3.2 Application Examples

A Vehicular Ad hoc Networks (VANETs) is a kind of mobile ad hoc networks where each vehicle serves as a node interconnected by wireless links. In this section, the PAA scheme for VANETs based on DSRC and IEEE 802.11p [2] is performed as follows. We will first describe the PHY format of IEEE 802.11p, followed by the performance results.

4.3.2.1 IEEE 802.11p Support

The IEEE 802.11p has constructed an OFDM-based PHY layer to operate in the 5.85–5.925 GHz unlicensed national information infrastructure band with 10 MHz bandwidth. Figure 4.4 shows the block diagram of IEEE 802.11p system model, where a 64-subcarrier OFDM system is employed. Among the 64 subcarriers, 52 are used for data transmission, which is further composed of 48 data subcarriers and 4 pilot subcarriers. The pilot signals are used for tracing the frequency offset and phase noise. Figure 4.5 shows the frame format which consists of the OFDM

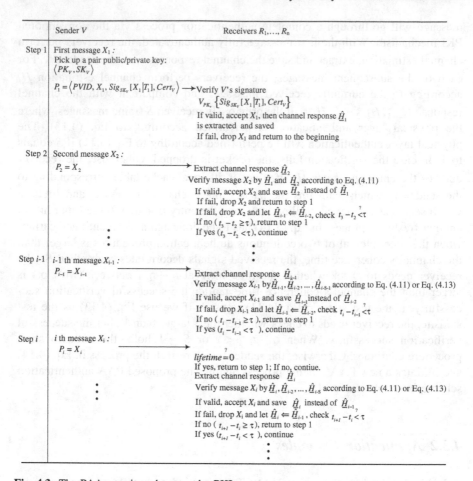

Fig. 4.3 The PAA security scheme under PKI

Physical Layer Convergence Protocol (PLCP) preamble, PLCP header, PLCP Service Data Unit (PSDU), tail bits, and pad bits. In the PLCP preamble field shown in Fig. 4.6, the preamble consists of 10 identical short training symbols and 2 identical long training symbols. The short training symbols and long training symbols, which are located in the preamble at the beginning of every PHY data packet, are used for signal detection; coarse frequency offset estimation, time synchronization, and channel estimation. A guard time GI, is attached to each data OFDM symbol in order to eliminate the Inter Symbol Interference introduced by the multi-path propagation. The proposed PAA scheme utilizes the two long training symbols pilot signal for channel estimation.

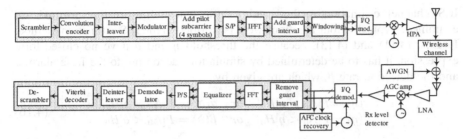

Fig. 4.4 System model of IEEE 802.11p physical layer

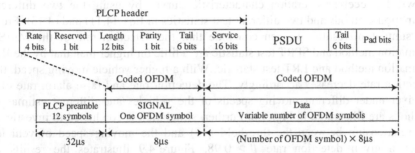

Fig. 4.5 The frame format of IEEE 802.11p (Transmission packet details)

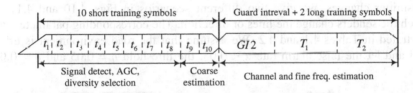

Fig. 4.6 Preamble structure of IEEE 802.11p

4.3.2.2 Performance Results

Simulation is conducted to verify the proposed PAA approach by collecting the PHY layer channel response from the two long training symbols in the pilot signal of each 802.11 frame. We set the simulation parameters as follows. There are $M = 52$ sub-carriers which consists of $N_d = 60$ OFDM symbols, bandwidth $B = 10$ MHz, and carrier frequency $f_c = 5.9$ GHz. The maximum delay spread is 200 ns corresponding to $M_\tau = 4$. The discrete Doppler spread M_v is determined as:

$$M_v = 2\left\lceil \frac{\omega_d}{B} \times M \times N_d \right\rceil \tag{4.15}$$

where ω_d is the maximum Doppler frequency in Hertz given by $\omega_d = f_c v_{\max}/c_0$, with $v_{\max}$ denoting the maximum terminal velocity, f_c being the carrier frequency, and c_0 denoting the speed of light. To estimate $\hat{\underline{H}}_k$ in Eqs. (4.11) and (4.12), both

ILS channel estimation method and MMSE channel estimation method are examined. Typically, four iterations were performed in the ILS estimation method. In Eqs. (4.11) and (4.13), because the threshold η and δ have no closed-form expression, it has to be determined by simulations according to the false alarm α and the detection rate β, which are given by

$$\begin{aligned} \alpha(\eta) &= P[\Lambda_0 > \eta | H_0] \quad or \quad \alpha(\delta) = P[\Lambda_0 > \delta | H_0] \\ \beta(\eta) &= P[\Lambda_0 < \eta | H_0] \quad or \quad \beta(\delta) = P[\Lambda_0 < \delta | H_0] \end{aligned} \tag{4.16}$$

In an urban road environment, we let false alarm rate $\alpha < 0.05$, the threshold $\eta = 0.05$ and $\delta = 0.05$, and the distance between vehicles be 25 m. Figure 4.7 shows the receiver operation characteristic curves by using the two different estimation methods and two different test statistics in Eqs. (4.11) and (4.13). From the simulation results, we can conclude that the detection rate of the MMSE estimation method and SPRT test statistic are a little bit higher than that by the ILS estimation method and LRT test statistic. With a higher vehicle moving speed, the detection rate decreases accordingly. The detection rate and false alarm rate of a receiver under different moving speeds of the vehicles and different estimation methods are illustrated in Fig. 4.8. Further, simulation is conducted to investigate the relationship between the threshold $\eta(\delta)$ and the moving speed of vehicles under a given detection rate $\beta \geq 0.98$. Figure 4.9 illustrates the results of threshold $\eta(\delta)$ as a function of the vehicles moving speed.

To check whether the proposed scheme can work well when the route of the transmitter changes, we give two different scenarios in Figs. 4.10 and 4.11, in which the senders change the lines or direction. The corresponding parameters are illustrated in Tables 4.1 and 4.2. We assume that the road environment is urban road and let the false alarm rate $\alpha < 0.05$, the threshold $\eta = 0.05$ and $\delta = 0.05$.

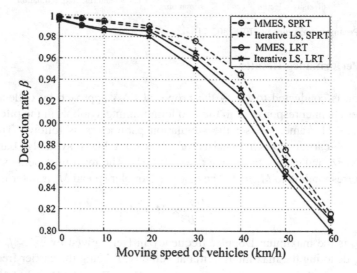

Fig. 4.7 The receiver operation characteristic curves

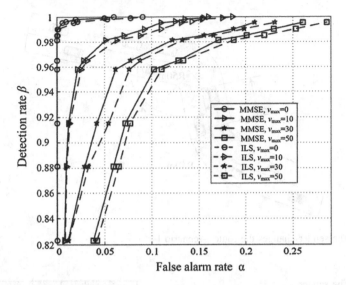

Fig. 4.8 The detection rate and false alarm rate of the receiver under different moving speed of the vehicles and different estimation methods

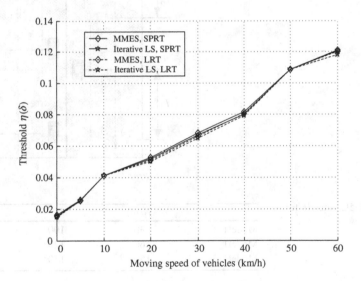

Fig. 4.9 The threshold as a function of the vehicles moving speed

The MMSE channel estimation method is used. The detection rate of the receivers under different moving speeds of the vehicles and different test statistic methods are illustrated in Figs. 4.12 and 4.13. From the simulation results, we can conclude that the detection rate of the receivers drops with moving speed increasing and the receivers yields the lower detection rate in scenario II than in scenario I.

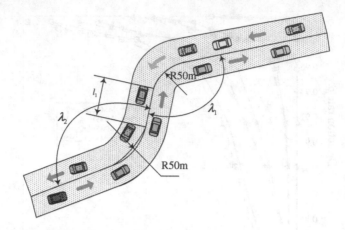

Fig. 4.10 The sender changes the route—scenario I

Fig. 4.11 The sender
changes the route—scenario
II

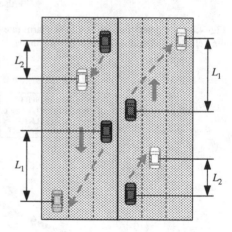

Table 4.1 Parameters for
Fig. 3.10

	$l_1(m)$	λ_1	λ_2
Model1_1	30	100°	100°
Model1_2	60	120°	120°
Model1_3	150	150°	150°

Table 4.2 Parameters for
Fig. 3.11

	$L_1(m)$	$L_2(m)$
Model2_1	50	100
Model2_2	100	200
Model2_3	150	300

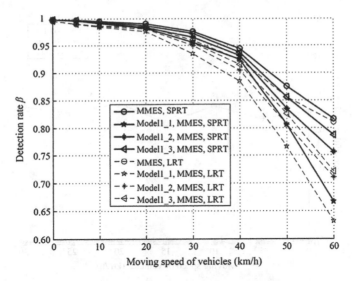

Fig. 4.12 The detection rate of the receivers under different moving speed of the vehicles and different test statistic methods when the sender changes route—scenario I

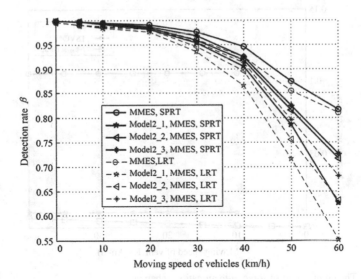

Fig. 4.13 The detection rate of the receivers under different moving speed of the vehicles and different test statistic methods when the sender changes route—scenario II

The current DSRC protocol is working at the 5.9 GHz in which the wavelength is 0.05 m. Because of the spatial uniqueness of Physical layer (PHY) channel responses, the channel response will change if the distance between two senders is larger than 0.05 m. Figure 4.14 illustrated the attacking results with different distance between Eve and the sender (in Fig. 4.1).

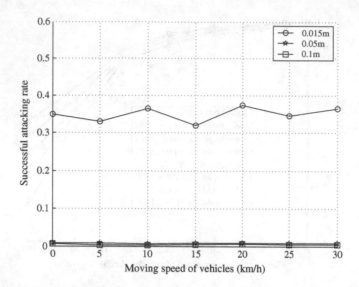

Fig. 4.14 The attacking rate with different distance between Eve and the sender

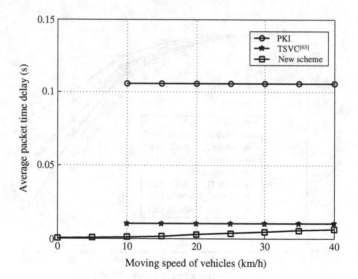

Fig. 4.15 The comparison of three authentication schemes

The comparison of several authentication schemes is shown in Fig. 4.15, which shows that the proposed PAA scheme yields the lowest time delay for authenticating each message. Note that the simulation result was derived by jointly considering the false positive and detection mismatch due to unexpected channel fluctuation.

4.4 Physical Layer Assisted Authentication Based on SCA

With the PAA scheme based on PKI, the sender initiates user authentication at the start of a session in order to establish the trust connection via the conventional PKI-based digital signature scheme. However, digital signatures are too expensive to compute on some resource limited application environment. Therefore, we need to construct our protocols from Symmetric Cryptography Algorithm (SCA). The computation costs of symmetric cryptography are low. Here we use Cipher Block Chaining Message Authentication Code [70] (CBC_MAC) as our initial security authentication Message Authentication Code (MAC) for PAA scheme.

4.4.1 Physical Layer Assisted Authentication Principle

Let the base station be a necessary part of our trusted computing base, which can serve as a key distribution server (KDS). At creation time, each mobile terminal is given a master key which is shared with the base station. All other keys are derived from this key. We launch and bootstrap the security with a shared key between each mobile terminal and the base station. By the Kerberos key agreement protocol [13], the base station is used as a trusted agent for key setup to achieve secure key agreement between mobile terminals.

To achieve two mobiles authentication and data integrity, a Message Authentication Code (MAC) need to be used. The message authentication codes are generated from a block cipher. The message is encrypted with some block cipher algorithm in Cipher Block Chaining (CBC) mode to create a chain of blocks such that each block depends on the proper encryption of the block before it. This interdependence ensures that a change to any of the plaintext bits will cause the final encrypted block to change in a way that cannot be predicted or counteracted without knowing the key to the block cipher. To calculate the CBC-MAC of message one encrypts in CBC mode with zero Initialization Vector (IV). A block diagram for computing CBC_MAC using a secret key K and a block cipher E is shown in Fig. 4.16.

Fig. 4.16 A block diagram for computing CBC_MAC

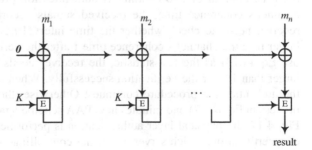

Alice is going to send the message to Bob, Bob need to verify the message from Alice, who generates nonce N_A randomly and sends it along with a request message R_A to Bob. Bob return nonce N_B with a response message R_B to Alice. Let K_{A-B} be the session key between Alice and Bob. The encrypted message has the following format: $M_E = [M]_{\langle K_{A-B} \rangle}$, where M is the message and the encryption key is K_{A-B}. So the protocol to provide freshness is:

$$
\begin{aligned}
Alice \rightarrow Bob &: \quad \{N_A, R_A\} \\
Bob \rightarrow Alice &: \quad \left\{ [R_B]_{<K_{A-B}>}, \quad MAC\left(K_{mac}, \quad N_A \big| T \big| [R_B]_{<K_{A-B}>}\right)\right\}
\end{aligned}
\tag{4.17}
$$

where T is timestamp for preventing replaying old messages; | is the concatenation operator. Whenever receiving any packet, the receivers first check if the timestamp found in packet is reasonable, and if so, continue. Otherwise, the receivers drop the packet since the receivers could be subject to replay attack. If the message authentication code (MAC) verify correctly, Alice knows that Bob generated the response after it sent the request. Let the message packets sent by Alice be denoted as $X_1, X_2, \cdots, X_N$, and X_i is transmitting packet. The first message packet X_1 that Alice sends to Bob is:

$$
Alice \rightarrow Bob: \quad \left\{ [X_1]_{<K_{A-B},T>}, \quad MAC\left(K_{mac}, \quad T \big| [X_1]_{<K_{A-B},T>}\right)\right\}
\tag{4.18}
$$

In Eqs. (4.17) and (4.18) the MACs are generated by CBC_MAC method shown in Fig. 4.16. The session key K_{A-B} and MAC key K_{mac} are derived from the master key by the Kerberos key agreement protocol [13]. When Alice sends the following message packets, she directly picks up the packet X_i without any authentication tag.

The receivers of the first message packet have to verify the MAC. After the MACs are verified correctly, the receivers perform channel measure, extract and save the channel responses $\hat{H}_1$. For each of the subsequent messages, the receivers measure the channel response $\hat{H}_i$ and compare with the previous channel response $\hat{H}_{i-1}, \hat{H}_{i-2}, \cdots, \hat{H}_{i-S}$, where the physical layer authentication is performed according to Eq. (4.11) or (4.13). If the message verifies successfully, the receiver saves $\hat{H}_i$ instead of $\hat{H}_{i-1}$ in order to authenticate the later packet. In case the verification fails, the receiver drops the message. As mentioned earlier, when the time interval of two continuous authentication procedures is larger than the channel's coherence time, the received signals decorrelates each other. So the receiver needs to check whether the time interval $t_{i+1} - t_i$ between two slots is larger than the channel's coherence time τ after the verification successfully. If we use Eq. (4.13) as the test statistic, the receiver needs to check whether the S is larger than 1 after the verification successfully. When $t_{i+1} - t_i < \tau$ or $S \geq 1$ hold, the authentication procedure continues. Otherwise, the sender has to refresh the process in Eq. (4.17) and initiates new PAA scheme again. From step 3 to step 5 in Fig. 4.17, the physical layer authentication is performed. We also set the lifetime of a certification, which serves as a timer controlling how long the certification is

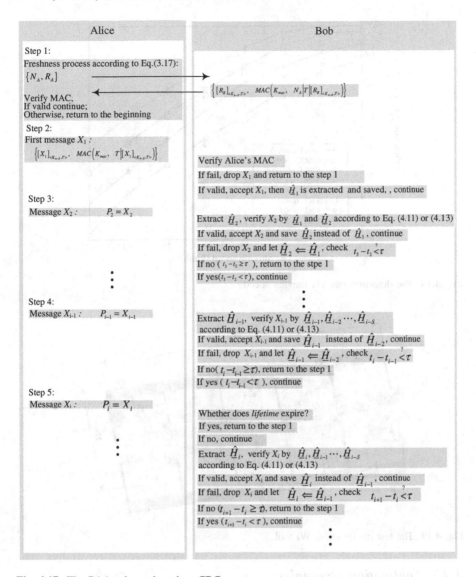

Fig. 4.17 The PAA scheme based on CBC

active. If the timer hits 0, the certification expired; another freshness process in Eq. (4.17) has to be done again. The proposed PAA authentication scheme is shown in Fig. 4.17.

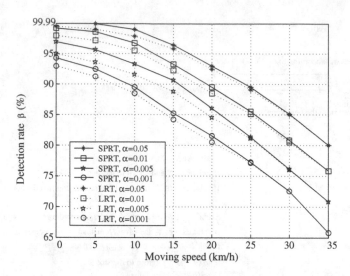

Fig. 4.18 The detection rate via moving speed

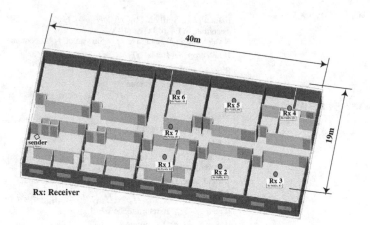

Fig. 4.19 The test model under WI tool

4.4.2 Application Examples

We apply CBC based PAA schemes to IEEE 802.11a [75] OFDM system to illustrate their performance via numerical simulations. We also model "typical" channel responses and to capture the spatial, frequency and time variability to these responses by software Wireless InSite (WI) tool, which is the software provides a broad range of site-specific predictions of propagation and communication channel characteristics in complex urban, indoor, rural and mixed path environments. Then we compare the result channel responses from software modeling with the channel responses estimated by some classic channel estimation methods (i.e., ILS and

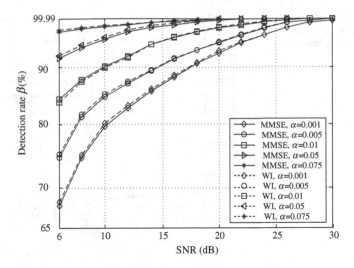

Fig. 4.20 The detection rate β under different false alarm α

Fig. 4.21 Probability of successful authentication under different SNR

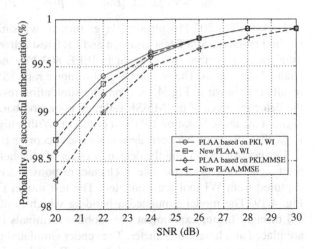

MMSE channel estimation methods), in which simulations are conducted to verify the proposed PAA approaches by collecting the PHY layer channel response from the two long training symbols in the pilot signal of each 802.11 frame. We set the simulation parameters as follows. There are $M = 52$ sub-carriers which consists of $N_d = 60$ OFDM symbols, bandwidth $W = 20$ MHz, and carrier frequency $f_c = 5.0$ GHz. In Eqs. (4.11) and (4.13), because the threshold η_0 and δ_0 have no closed-form expression, they have to be determined by simulations according to the false alarm α and the detection rate β, which are given by

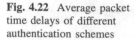

Fig. 4.22 Average packet time delays of different authentication schemes

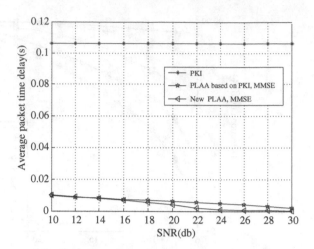

$$\alpha(\eta_0) = P[\Lambda_0 > \eta_0 | H_0] \quad or \quad \alpha(\delta_0) = P[\Lambda_0 > \delta_0 | H_0]$$
$$\beta(\eta_0) = P[\Lambda_0 < \eta_0 | H_0] \quad or \quad \beta(\delta_0) = P[\Lambda_0 < \delta_0 | H_0]$$
$$(4.19)$$

Under IEEE 802.11p physical layer format, we simulated MMSE estimation method based on SPRT test criterion and LRT test criterion, respectively. The two long training symbols pilot signal in IEEE 802.11p the packet are employed as channel estimation. The channel Signal-to-noise ratio (SNR) is fixed to 30 dB. The results are shown in Fig. 4.18. From the simulation results, we can conclude that the detection rates of the MMSE estimation method based on SPRT are a little bit higher than those based on LRT test criterion. With a higher vehicle moving speed v_{max}, the difference between them decreases accordingly.

Further, for examining the detection effect of channel responses estimated from pilot signals, the detection rate of channel responses derived from pilot signal and captured from WI tool are evaluated. The test model under WI tool is shown in Fig. 4.19. The model simulated a building which is 40 m long, 19 m wide and 3.5 m high. In our experiment, the mobiles terminals (i.e., sender and receivers) are placed at a height of 1 meter. The sender simulated the IEEE 802.11a physical layer packets and sent these packets out. By WI tools we capture the channel responses, which is compare with the channel responses that are estimated by method in [52]. The results are shown in Fig. 4.20, where the detection rate is given under different false alarm and SNR with no moving speed. Further, we let the false alarm $\alpha = 0.01$ and investigate the probability of successful authentication of new schemes under different SNR.

The performance of two different kinds of PAA scheme is compared. As shown in Fig. 4.21, the scheme based PKI and the proposed scheme based CBC have very close authentication probabilities. The based CBC scheme has a lighter overhead. Here the AES algorithm Rijndael [74] is chosen as encryption function in CBC_MAC based scheme. We choose the key size as 128 bits. The comparison of the average packet time delays for different authentication schemes is shown in

Fig. 4.22, which shows that the proposed novel schemes yield the lower time delay for authentication each message.

4.5 Summary

PHY authentication can provide fast and light weight authentication. All the previously reported PHY authentication approaches require launching an authenticated data packet at the beginning to initiate a trust connection between the sender and receivers, and it needs to be performed whenever there is any new receiver joining the network. Furthermore, a recipient may loss the trust connection due to false-positive recognition of channel response, which may happen frequently in the scenario of high user mobility. To take the best advantage of the fast physical layer authentication mechanism while minimizing the overhead in the trust initiation, this chapter aims to create a novel cross-layer security framework that can support all the previously reported physical layer authentication schemes [45–48]. The proposed framework can be implemented using two different initial authentication mechanisms. Firstly, an efficient PKI-based digital signature scheme based on physical layer authentication is developed. The second scheme is a simpler version of PKI-based PAA, which takes advantage of CBC symmetric key authentication mechanisms for resource limited networking environment.

With PAA, the sender initiates user authentication at the start of a session in order to establish the trust connection. For this purpose, the first frame of the session will be authenticated at the recipient via the conventional PKI-based or CBC based digital signature scheme. At this moment, the recipients have to record the channel responses specific to the recipient. The PAA scheme builds a trust connection between the sender and receiver by comparing the channel response of the currently received message with the channel estimation of the previously received message. It is possible that a receiver loses the trust connection with the sender due to false positive. Similar to TSVC or any other Timed Efficient Stream Loss-tolerant Authentication (TESLA) based approach, in such a circumstance the receiver has to wait for the next message that can be fully authenticated from the sender in order to restore the trust connection.

The presented schemes aim at achieving fast authentication and minimizing the packet transmission overhead without compromising the security requirements. Introduced schemes have strong robustness. Because the channel response that the entity extracted depends on the position that the entity is located, an adversary who compromises any entity and extracts the sender's channel response will not compromise the security of the whole system. The entities which located on different positions will extract the different channel responses, which mean each entity has its own channel response as its individual key. When an entity leaves the group, it is not necessary to revoke the key process.

Chapter 5
Detection of Node Clone and Sybil Attack Based on CI Approach

A Wireless Sensor Networks (WSNs) is an interconnected system of a large set of physically small, low cost, low power sensors that provide ubiquitous sensing and computing capabilities. The sensors have the ability to sense the environment in various modalities, process the information, and disseminate data wirelessly. Therefore, if the ability of the WSNs is suitably harnessed, it is envisioned that the WSN can reduce or eliminate the need for human involvement in information gathering in broad civilian or military applications such as national security, health care, environment protection, energy preservation, food safety, traffic management (vehicular ad hoc networking) and so on.

For sensor nodes deployed in hostile environments, attackers can easily inject malicious data or alter the content of legitimate messages during multi-hop forwarding due to the nature of wireless communication in sensor networks. Therefore, WSNs are vulnerable to much threat, among which node clone attack is an awfully harmful one. An adversary can capture a few nodes, extracts code and all secret credentials, and uses those materials to clone many nodes out of off-the-shelf sensor hardware. Then those nodes that seem legitimate are able to join the network and cause severe damages. Another awful attack to WSNs is Sybil attack. In the Sybil attack, malicious nodes masquerade other nodes or claim the faked ID in WSN. In the worst case, attackers that successfully launch these two attacking methods may generate an arbitrary number of additional node identities, using only one physical device and can even launch DoS (Denial-of-service attack) attacks to legitimate nodes by reducing their share of the resource and giving the attackers more resource to perform other attacks [77, 78].

With the inherent resource limitations of WSNs devices, lightweight and effective detection method is needed. References [45–48] presented a physical layer authentication idea, which is based on a generalized channel response with both spatial and temporal variability, and consider correlations in the time, frequency and spatial domains. In this chapter, we follow this idea and propose a new lightweight method for the node clone and Sybil attack detection in WSNs, which is based on Channel Identification (CI). A Sybil node or clone node is distinguished by the channel response which is extracted and used to estimate the

H. Wen, *Physical Layer Approaches for Securing Wireless Communication Systems*, SpringerBriefs in Computer Science, DOI: 10.1007/978-1-4614-6510-2_5, © The Author(s) 2013

channel information between two nodes. Finally, by employing Wireless InSite (WI) tools [73] and NS-2 and MATLAB software we experimentally validated our scheme under different networks. In additional, our approach incurs much less transmitting data and light overhead.

5.1 Detection Node Clone Based on Channel Identification

There are three general ways to detect clone nodes. References [79, 80] proposed the use of location-based keys to thwart and defend against several attacks, including the node clone. However, those methods to depend on the accurate geographic location which is difficult to apply to WSNs since the hardware of WSNs nodes is inefficient. References [81–84] introduced the key identity validation method and the identity of entity validation approach, which employed cryptographic related method to prevent node clone attacks. However, WSNs may not possible to deploy cryptographic-based schemes because of the resource constraints on sensor nodes. Sheng et al. [85] presented the radio resource testing approach as a defense against node replication attack, which is based on the assumption that a radio can not send and receive simultaneously on more than one channel. Some WSNs such as Vehicular Ad hoc Networks (VANETs) can not apply this scheme since a node may cheaply acquire multiple radios.

The goal of detecting clone node attacks is to validate that each node identity is the only identity presented by the corresponding physical node. Clone nodes generated from the same malicious node locate in the same position. Therefore, the channel estimation results from them must "close" to each other. If two or more nodes have very "close" the channel estimation information, a conclusion can be drawn that these nodes locate at the same position and they are the clone nodes generated from the same malicious node.

5.1.1 Network Model and Security Goals

We consider a densely deployed sensor network consisting of N resource-constrained sensor nodes. It is assumed that all the nodes have the same transmission range and communicate via bidirectional wireless links. Sensor nodes operate without supervision at most of time, and they can function correctly in a dynamic network, where new nodes are added, or old nodes disappear. In addition to neighbor nodes, sensor nodes only know some of other distant nodes. Nodes do not realize the network geographic outline. The average degree of node, that is, the number of its neighbors, denoted by d, varies with networks.

There might be or not be a powerful base station in our modeled network, but there does exist a trusted role named initiator that is responsible for initiating a round of distributed detection. Otherwise, an adversary can readily launch a DoS attack to the system by keeping launching detection procedures and exhausting nodes energy. The initiator can be the base station if one exists, or can be selected among all nodes via a distributed leader election.

In this section, we deal with the node clone attack, that is, the attempt by an adversary to add one or more nodes to the sensor networks by cloning captured nodes. The proposed scheme should detect the occurrence of the attack with an overwhelming probability. In addition, the scheme should make genuine nodes to stop communicating with clones nodes, expelling all clones off the network. Our scheme is based on the assumption that the majority of the sensor nodes are working properly. If an inspector successfully finds a clone, it becomes a witness. As more witnesses improve the resilience against the adversary, the number of witnesses represents a major security measurement for the scheme.

Usually, energy is the most valuable resource in wireless sensor networks. Communication consumes at least one order of magnitude power than any of other operations. Therefore, we evaluate the communication overload as the main performance of the scheme. On the other hand, sensor nodes are equipped with a limited amount of memory; thus any schemes requiring high storage would be considered as impractical. The memory requirement would be another performance metric for efficiency.

5.1.2 Node Clone Detection Principle Based on CI

The receiver uses channel estimates in two frames, $\hat{\underline{H}}_{t_1}$ and $\hat{\underline{H}}_{t_2}$ from two communication partners respectively, which can be extracted by Eqs. (4.3)–(4.7), to determine whether they are located in the same place. In the null hypothesis H_0, the claimants are in the same position. Otherwise, in the alternative hypothesis H_1, the claimant nodes are in the different positions. The notation $\sim$ is used to denote accurate values without measurement error, and thus we have:

$$
\begin{aligned}
H_0 &: \quad \underline{\tilde{H}}_{t_1} \to \underline{\tilde{H}}_{t_2} \\
H_1 &: \quad \underline{\tilde{H}}_{t_1} \mapsto \underline{\tilde{H}}_{t_2}
\end{aligned}
\tag{5.1}
$$

Same as Eq. (4.11), we have the LRT statistic:

$$
\Lambda_{i,j}^k = \frac{K_{co}\left\|\hat{\underline{H}}_{t_1} - \hat{\underline{H}}_{t_2}\right\|^2}{\left\|\hat{\underline{H}}_{t_2}\right\|^2} \underset{H_0}{\overset{H_1}{\gtrless}} \eta
\tag{5.2}
$$

where $\hat{\underline{H}}_t$ represents the channel response with measurement errors; K_{co} is the normalization factor and let threshold $\eta \in [0, 1]$.

In our consideration, an adversary can compromise sensor nodes which release all their security information to the attacker. Thereafter, the adversary can start replicating the node, and distribute the clones throughout the network. Note that

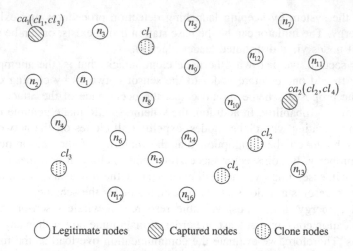

Fig. 5.1 WSNs under clone nodes

the cloned nodes own all the legitimate information of the compromised node (e.g. ID, keys, code, etc.). Thus the replicas can easily participate in the network operation in the same way as the legitimate nodes. The detection scheme depends on the richly scattered multipath environment. Because the received signal rapidly decorrelates over a distance of roughly half a wavelengths and the spatial separation of one to two wavelengths is sufficient for assuming independent fading paths. By collection the channel responses form all the nodes and comparison them with each other, we can check the clone nodes.

As shown in Fig. 5.1, points filled cycles are clone nodes generated from the captured nodes ca_i. Obviously, the position of clone nodes cl_j and the captured nodes ca_i are different, but they all have one identity (ID). For example, if nodes cl_1 and cl_3 are generated from the node ca_1, these three nodes are located at the different positions. The IDs of them are same. Let $H_t(cl_i \rightarrow n_i)$ denote the channel information between clone node cl_i and legitimate node n_i at time slot t. In the null hypothesis H_0, the claimants are in the same position. From Eq. (5.2), We have:

$$\Lambda_{i,j}^k = \frac{K_{co}\left\|\hat{\underline{H}}_{t_1}(cl_i \rightarrow n_k) - \hat{\underline{H}}_{t_2}(cl_j \rightarrow n_k)\right\|^2}{\left\|\hat{\underline{H}}_{t_2}(cl_j \rightarrow n_k)\right\|^2} \begin{array}{c} H_1 \\ \gtrless \\ H_0 \end{array} \eta \qquad (5.3)$$

If we have measured $H_{t_1}(cl_1 \rightarrow n_1)$ and $H_{t_2}(cl_3 \rightarrow n_1)$ such that $t_1 \leq t_2 + \tau$ and $\Lambda_{cl_1,cl_3}^{n_1} > \eta$ in (5.2), node n_1 can conclude that nodes cl_1 and cl_3 are in the different positions along with one ID and the clone nodes attack happened, where the time interval of two continuous signal pilots should be less than the channel's coherence time τ.

5.1.3 Determinant of Suspect Nodes

Node-to-node broadcasting is a quite practical, effective way to distributive detect the node clone. Every node collects all of its neighbor's identities along with their channel response, and broadcasts to the network. When a node receives a broadcasted message from others, it compares those nodes listed in the message with its own neighbors and revokes neighbor nodes that have collision channel responses. Our channel identification scheme relies on the estimation the channel information of periodical messages. The details of the scheme are presented as follows.

All the nodes are randomly divided into two groups: claimer group and witness group in which the nodes are denoted as n_i^c and n_i^w, respectively. The nodes would periodically play these two roles. The witness nodes periodically broadcast request messages R_w to claimer nodes with timestamp T_w. All neighboring claimer nodes, within the signal range of the witness nodes, will receive the previous request messages R_w. They return response messages R_c to the witness nodes within time interval $\Delta t + \Delta \Omega$, where Δt is response delay and $\Delta \Omega$ should be less than the channel's coherence time τ. The response message R_c can be in the following format:

$$R_c : \{Node\ ID,\ pilot\} \tag{5.4}$$

After witness nodes receive the messages R_c, they extract channel measure $\hat{\underline{H}}_{t_i}(n_i^c \rightarrow n_k^w)$, $i = 1, 2, \cdots, N_c$, $k = 1, 2, \cdots, N_w$, where N_c and N_w are number of nodes in claimer group and witness group, respectively. Then they perform comparison according to Eq. (2.2) and get $\Lambda_{i,j}^k$, $i \neq j \neq k$, $i, j = 1, 2, \cdots, N_c$, $k = 1, 2, \cdots, N_w$.

There are three scenarios as shown in Fig. 5.2. Figure 5.2a introduces the first scenarios, in which two or more malicious nodes surround a witness node. Nodes ca_1 and cl_1 are captured node and clone node generated from ca_1. The witness node is n_1. If $|t_i - t_j| \leq \Delta \Omega$ and $\Lambda_{i,j}^1 < \eta$ for nodes ca_1 and cl_1, n_1 can conclude that nodes ca_1 and cl_1 are suspect nodes. Then these suspect nodes are saved into $suslist_{n_1^w}$. Node n_1^w broadcast $suslist_{n_1^w}$ to the network and expel the clone node.

The second scenario is shown in Fig. 5.2b, in which two or more malicious nodes belong to different witness nodes and the witness nodes share part of their neighbor nodes. Figure 5.2c shows the final scenario, in which two or more malicious nodes belong to different witness nodes and the witness nodes are far from each other and do not share any of their neighbor nodes. In these two situations, witness nodes can not find suspect nodes by performance comparison according to Eq. (5.2). Therefore, every witness node broadcasts a list of its neighbor nodes to other witness nodes. After a witness node n_k^w received the lists of their neighbor nodes from other witness nodes, it checks all the lists and save the witness nodes ID that contain at least one same notes ID with it into same node ID list $snIDlist_{n_k^w}$. Then the proposed scheme has to depend on sending the random

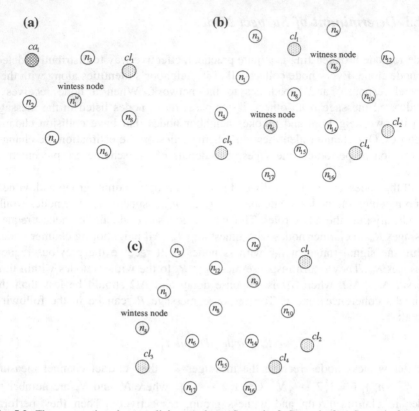

Fig. 5.2 Three scenarios about malicious nodes. **a** Scenario. **b** Scenario 2. **c** Scenario 3

sequence to find the suspect nodes. All the witness nodes in $snIDlist_{n_k^w}$ will send random choose sequences $sq_{n_k^w}$ which are different in different witness node to their neighbor nodes and ask all their neighbor nodes send back the received sequences along with their ID,

$$R_s : \{Node\ ID,\ sq_{n_k^w}\}. \tag{5.5}$$

Then every witness node broadcasts a $report_{n_k^w}$ that contain all their neighbor nodes' R_s to other witness nodes. The witness nodes check the $report_{n_k^w}$. If two nodes with same ID hold different random sequence, the clone nodes attack happened.

We still take Fig. 5.2 as an example. In Fig. 5.2a, Node n_1 is the witness node; and other seven nodes are its neighbor nodes, in which nodes ca_1 and cl_1 have same ID. Firstly, Node n_1 broadcasts request message R_w. After neighbor nodes received R_w, they return response messages to n_1, in which the responses of nodes

ca_1 and cl_1 are $R_c^{ca_1} = \{ca_1, pilot\}$ and $R_c^{cl_1} = \{cl_1, pilot\}$ within the response time interval, respectively. The node n_1 extracts channel response $\hat{\underline{H}}_{t_1}(ca_1 \rightarrow n_1)$ and $\hat{\underline{H}}_{t_2}(cl_1 \rightarrow n_1)$ from the *pilot* and calculates the $\Lambda_{ca_1, cl_1}^{n_1}$. If $\Lambda_{ca_1, cl_1}^{n_1}$ is less than threshold η, node n_1 can determine that nodes ca_1 and cl_1 are suspect nodes. Then ca_1 and cl_1 are saved into $suslist_{n_1}$. The node n_1 will broadcast the $suslist_{n_1}$ and expel the nodes ca_1 and cl_1.

Let us consider the situation in the Fig. 5.2b, we assume that the witness nodes are randomly chose as nodes n_6 and n_9. Their neighbor nodes are listed in Table 4.1. We can know that the clone nodes cl_1 and cl_3 belong to different witness node, and the witness nodes can not find them as the situation in Fig. 5.2b. Then the detection scheme has to depend on sending the random sequence to find the suspect nodes. Witness nodes n_6 and n_9 send random choose sequences sq_{n6} and sq_{n9} to their neighbor nodes, respectively and ask all their neighbor nodes send back the received sequences along with their ID $R_s : \{Node\ ID,\ sq_{n_k^w}\}$. Then witness nodes n_6 and n_9 broadcasts $report_{n6}$ and $report_{n9}$. After received $report_{n6}$ and $report_{n9}$, the witness nodes check the $report_{n6}$ and $report_{n9}$. Nodes cl_1 and cl_3 are generated from the captured node ca_1, therefore, their identity are the identity of the node ca_1. The witness nodes n_6 and n_9 can find that nodes cl_1 and cl_3 have same identity and different random sequence. It is concluded that the clone attack happened. We note that the nodes n_{14} and n_8 are not only the neighbor node of n_6 but also the neighbor node of n_9, therefore, they can receive both sq_{n_6} and sq_{n_9}.

It is obviously that the random sequence scheme will fail to detect the clone nodes if there is a node is the neighbor nodes of two or more witness nodes in the same time. For example, if the node n_8 is the node generated from the captured node ca_1, it can receive both sq_{n_6} and sq_{n_9}, it can transfer these two sequence to the nodes cl_1 and cl_3. As a results, the witness nodes n_6 and n_9 will get wrong conclusion that the nodes cl_1 and cl_3 are legitimate and are the neighbor nodes of them. For avoiding this problem, we modify the random sequence scheme as following. Instead of sending the random sequence to their neighbor nodes, the witness nodes will send the random sequence along with timestamp T_s. The response of the neighbor nodes become:

$$R_s : \{Node\ ID,\ sq_{n_k^w}, T_s\}. \tag{5.6}$$

If a node send the received sequence transferred from other nodes, timestamp T_s will become $T_s + 1$. The proposed scheme is shown in Fig. 5.3 (Table 5.1).

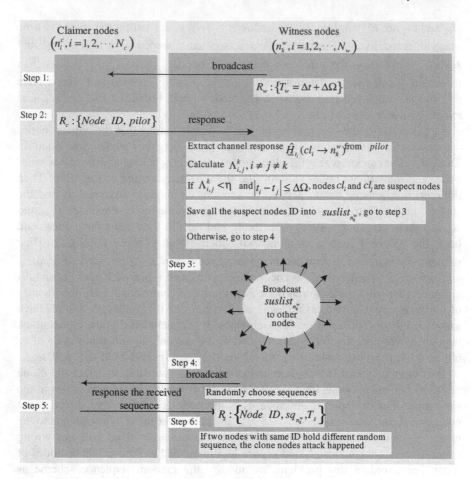

Fig. 5.3 Proposed scheme for detection clone nodes

Table 5.1 The neighbor nodes and their received sequence along with ID

	The neighbor nodes and its received sequence with ID						
n_6	cl_3	n_4	n_1	n_8	n_{14}	n_{15}	n_{17}
Identity	ID_{ca_1}	ID_{n_4}	ID_{n_1}	ID_{n_8}	$ID_{n_{14}}$	$ID_{n_{15}}$	$ID_{n_{17}}$
Received sequence	sq_{n_6}	sq_{n_6}	sq_{n_6}	sq_{n_6}, sq_{n_9}	sq_{n_6}, sq_{n_9}	sq_{n_6}	sq_{n_6}
n_9	cl_1	n_5	n_{12}	n_{10}	n_{14}	n_8	
Identity	ID_{ca_1}	ID_{n_5}	$ID_{n_{12}}$	$ID_{n_{10}}$	$ID_{n_{14}}$	ID_{n_8}	
Received sequence	sq_{n_9}	sq_{n_9}	sq_{n_9}	sq_{n_9}	sq_{n_9}, sq_{n_6}	sq_{n_9}, sq_{n_6}	

5.1.4 Performance Results

5.1.4.1 Simulation System Set up

We apply our novel scheme to IEEE 802.15.4 network to illustrate its performance via numerical simulations by MATLAB and NS-2 tools. We also model "typical" channel responses and to capture the spatial, frequency and time variability to these responses by software Wireless InSite (WI) [73] tools, which is the software provides a broad range of site-specific predictions of propagation and communication channel characteristics in complex urban, indoor, rural and mixed path environments. Then we compare the result channel responses from software modeling with the channel responses estimated by the classic channel estimation method (i.e., Iterative Least-Squares (ILS) channel estimation method introduced in [52]). We set the simulation parameters as follows. The carrier frequency is $f_c = 2.4$ GHz and bandwidth is 2 MHz. In Eq. (5.2), because the threshold η has no closed-form expression, it has to be determined by simulations according to the detection rate β and the false alarm α, which denotes the misdetection rate.

The simulation results will be based on the square area with 40 m long and 19 m wide. Further, for examining the detection effect of channel responses estimated from pilot signals, the detection rate of channel responses derived from pilot signal and captured from WI tool are evaluated. The test model under WI tool is shown in Fig. 5.4. The model simulated a building which is 40 m long, 19 m wide and 3.5 m high. In the network the clone nodes are called bad nodes. Each experimental result is calculated from an average over 500 independent simulations.

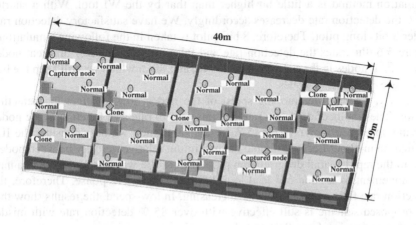

Fig. 5.4 The test model under WI tool

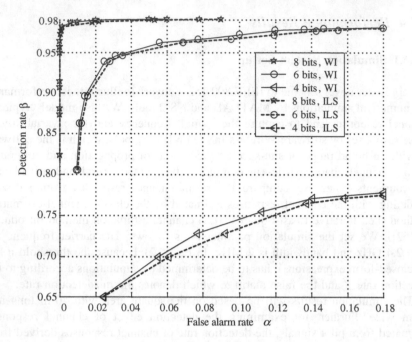

Fig. 5.5 Detection rate and false alarm rate under different number of pilot bits

5.1.4.2 Successful Probability of Detection

We let the threshold $\eta = 0.05$, and the number of all nodes is 40 in which the density of bad nodes is 20 %. In the following simulation, we all use these parameters if there is no special explanation. Figure 5.5 shows the detection rate and false alarm rate under the different number of pilot bits with no moving speed. From the simulation results, we can conclude that the detection rate of the ILS estimation method is a little bit higher than that by the WI tool. With a shorter pilot, the detection rate decreases accordingly. We have satisfactory detection rate under 8 bits long pilot. Therefore, 8 bits pilot is taken in the following simulations. Figure 5.6 illustrates the detection rate and false alarm rate under different nodes density. All nodes in the networks are 20, 40, and 80 in which we still keep the bad nodes are 20 % of all nodes.

Finally, we consider moving speed of the nodes how to affect the detection results. The results of detection rate and false alarm rate as a function of the nodes moving speed are shown in Fig. 5.7. In this simulation we only consider the ILS estimation method. Let v_{max} denote the maximum relative velocity between nodes. Due to the rapid spatial decorrelation properties of the wireless multipath channel, the movement of a node can lead to a different channel response. Therefore, the detection rate falls with the velocity increasing. In low speed, the results show that the proposed scheme is still effective with over 85 % detection rate with misdetection rate varying from 0.01 to 0.1.

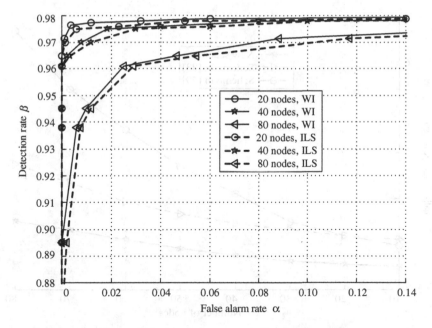

Fig. 5.6 Detection rate and false alarm rate under different nodes density

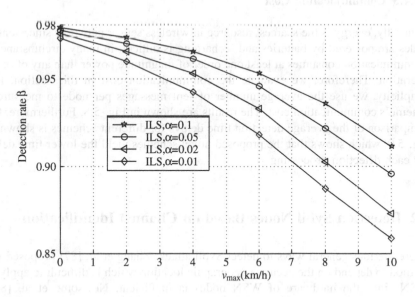

Fig. 5.7 Detection rate under different nodes moving speed

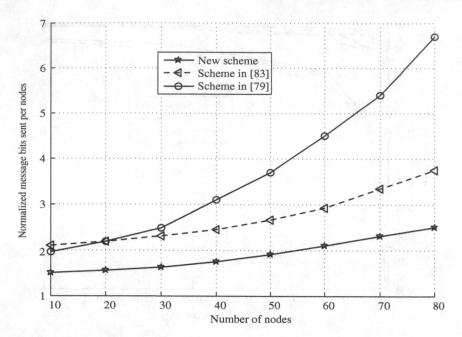

Fig. 5.8 Communication cost

5.1.4.3 Communication Cost

Generally, energy is the scarcest resource in wireless sensor networks since sensor nodes are powered by batteries and recharging is difficult in many circumstances. Communication consumes at least one order of magnitude power than any of other operation. Therefore, we evaluate the communication cost in this section. For simplicity, we use the average number of sent messages per node to measure a scheme's communication cost. The results are shown in Fig. 5.8. Furthermore, the comparison of the average detection time delays for different schemes is shown in Fig. 5.9, which shows that the proposed novel schemes yield the lower time delay for each detection processing.

5.2 Detection Sybil Nodes Based on Channel Identification

There are four general ways to detect Sybil attack. Zhang et al. [87] proposed the method to depend on the accurate geographic location which is difficult to apply to WSN since the hardware of WSN nodes is inefficient. Newsome et al. [88], Eschenauer and Gligor [89] and Golle et al. [90] introduced the key identity validation method and the identity of entity validation approach, which employed cryptographic related method to prevent Sybil attacks. However, WSN may not

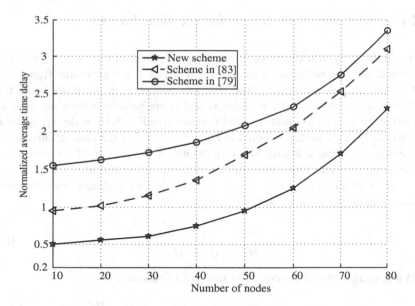

Fig. 5.9 Average detection time delays of different schemes

possible to deploy cryptographic-based schemes because of the resource constraints on sensor nodes. Newsome et al. [88] presented the radio resource testing approach as a defense against Sybil attack, which is based on the assumption that a radio can not send and receive simultaneously on more than one channel. Some WSNs such as Vehicular Ad hoc Networks (VANETs) [90] can not apply this scheme since a node may cheaply acquire multiple radios. A number of position verification techniques [91–92] have been proposed. Those approaches have to depend on the additional hardware except Received Signal Strength Indication (RSSI) method. However, it is difficult for RSSI to measure the precise distance and signal strength because of the influence of reflection, multi-path fading and none line of sight in the wireless communication. For many applications, successful Sybil attacks may be expensive even when the Sybil attack cannot be prevented.

In the Sybil attack, malicious nodes masquerade other nodes or claim the faked ID in WSNs. The goal of detecting Sybil attacks is to validate that each node identity is the only identity presented by the corresponding physical node. Sybil nodes generated from the same malicious node locate in the same position. Therefore, the channel estimation results from them must "close" to each other. If two or more nodes have very "close" channel estimation information, a conclusion can be drawn that these nodes locate at the same position and they are the Sybil nodes generated from the same malicious node.

5.2.1 Sybil Nodes Detection Principle Based on CI

Scenario under Sybil nodes is shown in Fig. 5.10. Dashed cycles are Sybil nodes generated from the malicious node, in which s_1 and s_2 are generated from m_1; s_3 and s_4 are generated from m_2. The true position of s_i is the position of their malicious node. For example, if nodes s_1 and s_2 are generated from the node m_1, instead of the positions of nodes s_1 and s_2 claimed in Fig. 5.10, nodes s_1, s_2 and m_1 are physically at the same position. Let $H_t(s_i \rightarrow n_i)$ denote the channel information between nodes s_i and n_i at time slot t. In the null hypothesis H_0, the claimants are in the same position. Otherwise, in the alternative hypothesis H_1, the claimant nodes are in different positions. The notation $\sim$ is used to denote accurate values without measurement error, and thus have:

$$H_0: \quad \underline{\tilde{H}}_{t_1} \rightarrow \underline{\tilde{H}}_{t_2}$$
$$H_1: \quad \underline{\tilde{H}}_{t_1} \mapsto \underline{\tilde{H}}_{t_2} \tag{5.7}$$

We normalize the likelihood ratio test (LRT) statistic as:

$$\Lambda_{i,j}^k = \frac{K_{co}\left\|\hat{\underline{H}}_{t_1}(s_i \rightarrow n_k) - \hat{\underline{H}}_{t_2}(s_j \rightarrow n_k)\right\|^2}{\left\|\hat{\underline{H}}_{t_2}(s_j \rightarrow n_k)\right\|^2} \mathop{\gtrless}_{H_0}^{H_1} \eta \tag{5.8}$$

where $\hat{\underline{H}}_t$ represents the channel response with measurement errors; K_{co} is the normalization factor and let threshold $\eta \in [0, 1]$.

The detection scheme depends on the richly scattered multipath environment. Because the received signal rapidly decorrelates over a distance of roughly half a wavelengths and the spatial separation of one to two wavelengths is sufficient for assuming independent fading paths, the time interval of two continuous signal pilots should be less than the channel's coherence time τ. If we have measured

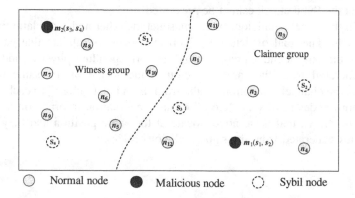

Fig. 5.10 WSNs under Sybil nodes

$H_{t_1}(s_1 \rightarrow n_1)$ and $H_{t_2}(s_2 \rightarrow n_1)$ such that $t_1 \leq t_2 + \tau$ and $\Lambda^{n_1}_{s_1,s_2} < \eta$ in (4.8), node n_1 can conclude that nodes s_1 and s_2 are in the same position and the Sybil attack happened.

5.2.2 Determinant of Suspect Nodes

Our channel identification scheme relies on the channel information estimation of periodical messages. The details of the proposed scheme are presented as follows. All the nodes are randomly divided into two groups: claimer group and witness group in which the nodes are denoted as n^c_i and n^w_i, respectively. The nodes would periodically play these two roles. The witness nodes periodically broadcast request messages R_w to claimer nodes with timestamp T_w. All neighboring claimer nodes, within the signal range of the witness nodes, will receive the previous request messages R_W. They return response messages R_c to the witness nodes within time intervals $\Delta t + \Delta \Omega$, where Δt is response delay and $\Delta \Omega$ should be less than the channel's coherence time τ. The response message R_c can be in the following format:

$$R_c : \{Node \ ID, pilot\} \tag{5.9}$$

After witness nodes receive the messages R_c, they extract channel measure $\hat{H}_{t_i}(n^c_i \rightarrow n^w_k)$, $i = 1, 2, \cdots, N_c$, $k = 1, 2, \cdots, N_w$, where N_c and N_w are number of nodes in claimer group and witness group, respectively. Then they perform comparison according to Eq. (5.8) and get $\Lambda^k_{i,j}$, $i \neq j \neq k$, $i, j = 1, 2, \cdots, N_c$, $k = 1, 2, \cdots, N_w$. If $|t_i - t_j| \leq \Delta \Omega$ and $\Lambda^k_{i,j} < \eta$ for nodes n^c_i and n^c_j, n^w_k can conclude that nodes n^c_i and n^c_j are suspect nodes. Then these suspect nodes are saved into $suslist_{n^w_k}$. Node n^w_k broadcast $suslist_{n^w_k}$ to other witness nodes. Our scheme is based on the assumption that the majority of the sensor nodes are working properly. If majority of witness nodes vote that nodes n^c_i and n^c_j are suspect nodes, the Sybil attack happened. The proposed scheme is shown in Fig. 5.11.

We still take Fig. 5.10 as an example. Node n_6 is the witness node; and nodes s_3 and m_1 are claimer nodes. Firstly, Node n_6 broadcasts request message R_w. After nodes s_3 and m_1 receive R_w, they return response messages $R^{s_3}_c = \{s_3, pilot\}$ and $R^{m_1}_c = \{m_1, pilot\}$ within the response time interval, respectively. According to Eqs. (4.3)–(4.7), node n_6 extracts channel response $\hat{H}_{t_1}(s_3 \rightarrow n_6)$ and $\hat{H}_{t_2}(m_1 \rightarrow n_6)$ from the pilot and calculates the $\Lambda^{n_6}_{s_3,m_1}$. If $\Lambda^{n_6}_{s_3,m_1}$ is less than threshold η, node n_6 can determine that nodes s_3 and m_1 are suspect nodes. Then s_3 and m_1 are saved into $suslist_{n_6}$ and wait for final decision. We assume that nodes $n_5, n_6, n_8, n_{10}, s_1$ are witness nodes. For nodes s_3 and m_1, final decision made by voting is shown in Table 5.2. Our scheme is based on the assumption that the majority of the sensor nodes is working properly. If the majority of witness nodes reports that a pair of nodes are Sybil nodes, the decision is made as Sybil attack happened. The scheme

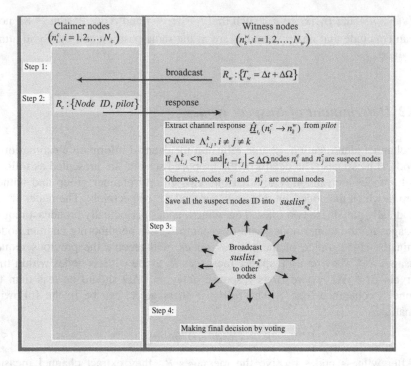

Fig. 5.11 Proposed scheme for detection Sybil nodes

Table 5.2 An example of voting

	n_5	n_6	n_8	n_{10}	s_1	Decision results
s_3, m_1	$\Lambda_{s_3,m_1}^{n_5} < \eta$	$\Lambda_{s_3,m_1}^{n_6} < \eta$	$\Lambda_{s_3,m_1}^{n_8} > \eta$ (detection error)	$\Lambda_{s_3,m_1}^{n_{10}} < \eta$	$\Lambda_{s_3,m_1}^{s_1} > \eta$ (false report)	Sybil attack happened
Voting	Sybil	Sybil	No Sybil	Sybil	No Sybil	

will make genuine nodes to stop communicating with Sybil nodes, expelling them off the network. From Table 5.2, we know that nodes s_3 and m_1 are detected as Sybil nodes and will be expelled off the network.

5.2.3 Performance Results

We apply our novel scheme to IEEE 802.15.4 network to illustrate its performance via numerical simulations by MATLAB and NS-2 tools. We also model "typical" channel responses and to capture the spatial, frequency and time variability to these responses by software Wireless InSite (WI) [73] tools, which is the software provides a broad range of site-specific predictions of propagation and communication channel characteristics in complex urban, indoor, rural and mixed path environments. Then we compare the result channel responses from software modeling with the channel responses estimated by the classic channel estimation method (i.e., ILS channel estimation method). We set the simulation parameters as follows. The carrier frequency is $f_c = 2.4$ GHz and bandwidth is 2 MHz. In Eq. (5.8), because the threshold η has no closed-form expression, it has to be determined by simulations according to the detection rate β and the false alarm α, which denotes the misdetection rate.

The simulation results will be based on the square area with 40 m long and 19 m wide. Further, for examining the detection effect of channel responses estimated from pilot signals, the detection rate of channel responses derived from pilot signal and captured from WI tool are evaluated. The test model under WI tool is shown in Fig. 5.12. The model simulated a building which is 40 m long, 19 m wide and 3.5 m high. In the network the Sybil nodes and malicious nodes are called bad nodes. Each experimental result is calculated from an average over 500 independent simulations.

We let the threshold $\eta = 0.05$, and the number of all nodes is 40 in which the density of bad nodes is 20 %. In the following simulation, we all use these parameters if there is no special explanation. Figure 5.13 shows the detection rate and false alarm rate under different the number of pilot bits with no moving speed. From the simulation results, we can conclude that the detection rate of the ILS estimation method is a little bit higher than that by the WI tool. With a shorter pilot, the detection rate decreases accordingly. We have satisfactory detection rate under 8 bits long pilot. Therefore, 8 bits pilot is taken in the following simulations. Figure 5.14 illustrates the detection rate and false alarm rate under different nodes density. All nodes in the networks are 20, 40, and 80 in which we still keep the bad nodes is 20 % of all nodes.

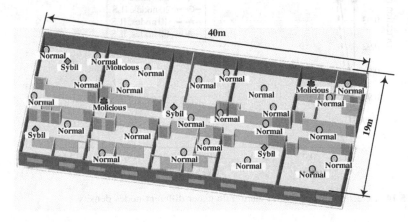

Fig. 5.12 The test model under WI tool

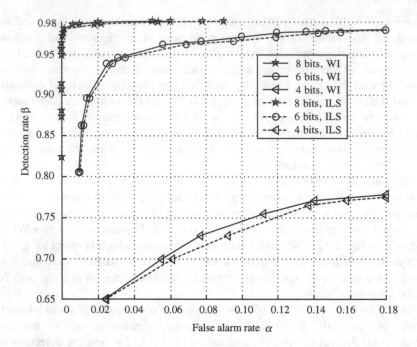

Fig. 5.13 Detection rate and false alarm rate under different number of pilot bits

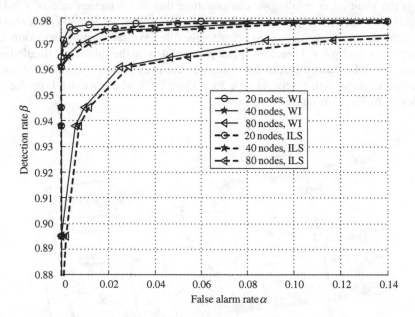

Fig. 5.14 Detection rate and false alarm rate under different nodes density

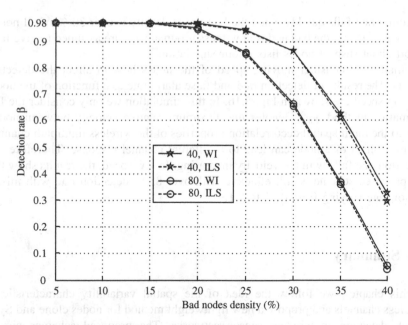

Fig. 5.15 Detection rate under different bad nodes density

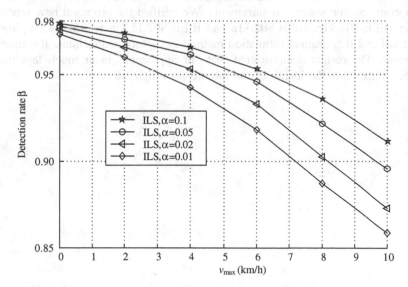

Fig. 5.16 Detection rate under different nodes moving speed

Further, simulation is conducted to investigate the relationship between the detection rate and bad nodes density under the false alarm rate $\alpha < 0.02$. The bad nodes density in the network varies from 5 to 40 %. Figure 5.15 illustrates the detection rate under different bad nodes density. It can be easily found that the detection rate falls quickly when the bad nodes density is over 30 % of all nodes. The reason is that the bad nodes in witness group always are assumed to give false report about suspect nodes lists in our simulation.

Finally, we consider moving speed of the nodes how to affect the detection results. The results of detection rate and false alarm rate as a function of the nodes moving speed are shown in Fig. 5.16. In this simulation we only consider the ILS estimation method. v_{max} denotes the maximum relative velocity between nodes. Due to the rapid spatial decorrelation properties of the wireless multipath channel, the movement of a node can lead to a different channel response. Therefore, the detection rate falls with the velocity increasing. In low speed, the results show that the proposed scheme is still effective with over 85 % detection rate with misdetection rate varying from 0.01 to 0.1.

5.3 Summary

In this chapter, we follow the idea of the spatial variability characteristic of wireless channels and propose a new lightweight method for nodes clone and Sybil attack detection in wireless sensor networks. The proposed schemes aim at achieving fast detection and minimizing the packet transmission overhead without compromising the security requirements. We verified the proposed new schemes under IEEE 802.11p, IEEE 802.11a and IEEE 802.15.4 networks by employing WI tools and ILS channel estimation method to capture and estimate the channel response. The results demonstrated that our approaches incur much less transmitting data while the identification rate keeps a high rate.

References

1. C. E. Shannon, "Communication theory of secrecy systems," Bell Syst. Tech. J., vol. 29, pp. 656–715, 1949.
2. A. D. Wyner, "The wire-tap channel," Bell Syst. Tech. J., vol. 54, pp. 1355–1387, Oct. 1975.
3. I. Csiszar and J. Korner, "Broadcast channels with confidential messages," IEEE Trans. Inf. Theory, vol. 24, pp. 339–348, May 1978.
4. A.O. Hero, "Secure space-time communication," IEEE Trans. Inf. Theory, vol. 49, no. 12, pp. 3235–3249, Dec. 2003.
5. X. Li and J. Hwu, "Using antenna array redundancy and channel diversity for secure wireless transmissions," J. Commun., vol. 2, no. 3, pp. 24–32, May 2007.
6. H. Kim and J. D. Villasenor, "Secure MIMO communications in a system with equal numbers of transmit and receive antennas," IEEE Commun. Lett., vol. 12, no. 5, pp. 386–388, May, 2008.
7. U. Maurer, "Secret key agreement by public discussion from common information," IEEE Trans. Inf. Theory, vol. 39, no. 3, pp. 733–742, Mar. 1993.
8. S. Wolf, Theoretically and computationally secure key agreement in cryptography, Ph. D Dissertation, 1999.
9. I. Csiszar and P. Narayan, "Common randomness in information theory and cryptography, part I: Secret sharing," IEEE Trans. Inf. Theory, vol. 39, pp. 1121–1132, July 1993.
10. A. Thangaraj, S. Dihidar, A. R. Calderbank, S. McLaughlin, and J.-M. Merolla, "Applications of LDPC codes to the wiretap channel," IEEE Trans. Inf. Theory, vol. 53, no. 8, pp. 2933–2945, Aug. 2007.
11. M. Nloch, J. Barros and M. R. D. Rodrigues, "Wireless information theoretic security," IEEE Trans. Inform. Theory, vol. 54, no. 6, pp. 2515–2534, June, 2008.
12. J. Muramatsu, "Secret key agreement from correlated source outputs using low density parity check matrices," IEICE Trans. Fund. Elec. Comm. Comp., vol. E89-A, no. 7, pp. 2036–2046, July 2006.
13. M. Yuksel and E. Erkip, "The relay channel with a wire-tapper," in Proceedings of 41st Annual Conference on Information Sciences and Systems, Baltimore, MD, Mar. 2007.
14. E. Tekin and A. Yener, "The general Gaussian multiple-access and two-way wire-tap channels: Achievable rates and cooperative jamming," IEEE Trans. Inf. Theory, vol. 54, no. 6, pp. 2735–2751, June, 2008.
15. L. Lai, H. El Gamal and H. V. Poor, "The Wiretap channel with feedback: encryption over the channel," http://www.ece.osu.edu/helgamal/publications.html.

H. Wen, *Physical Layer Approaches for Securing Wireless Communication Systems*, SpringerBriefs in Computer Science, DOI: 10.1007/978-1-4614-6510-2, © The Author(s) 2013

16. C. Berrou, A. Glavieux, and P. Thitimajshima, "Near Shannon limit error correcting coding and decoding: Turbo codes," in Proceedings of IEEE Int. Conf. Communications, Geneva, Switzerland, 1993, pp. 1064–1070.
17. R. G. Gallager, "Low density parity check codes," Cambridge, MA: MIT Press, 1963.
18. D. J. C. MacKay, "Good error-correcting codes based on very sparse matrices," IEEE Trans. Inf. Theory, vol. 45, no. 3, pp. 399–431, Mar. 1999.
19. L. Bazzi, T. J. Richardson and R. L. Urbanke, "Exact thresholds and optimal codes for the binary-symmetric channel and gallager's decoding algorithm A," IEEE Trans. Inf. Theory, vol. 50, no. 9, pp. 2010–2021, Sep. 2004.
20. B. Marco, C. Giovanni, and C. Franco, "Variable rate LDPC codes for wireless applications," in Proceedings of Software in Telecommunications and Computer Networks International Conference on Sept. 29–Oct. 1, pp. 301–305, 2006.
21. H. Mahdavifar and A. Vardy, "Achieving the secrecy capacity of wiretap channels using polar codes," IEEE Trans. Inform. Theory, vol. 57, no. 10, pp. 6428–6443, 2011.
22. A. Subramanian, A. Thangaraj, M. Bloch, and S. McLaughlin, "Strong secrecy on the binary erasure wiretap channel using large-girth ldpc codes," IEEE Trans. Inform. Forensics and Secur., vol. 6, no.3, pp. 585–594, 2011.
23. H. Ahmadi, R. Safavi-Naini, "Secret keys from channel noise," Eurocry 2011, pp. 266–283, 2011.
24. K. Zeng and D. Wu, A. Chan and P. Mohapatra, "Exploiting multiple-antenna diversity for shared secret key generation in wireless networks," IEEEINFOCOM 2010, pp. 1–9, 2010.
25. H. Behairy, S.-C. Chang, "Parallel concatenated gallager codes," Electron.Lett., vol. 36, no. 24, pp. 2025–2026, 2000.
26. H. Wen, G. Gong and P-H. Ho, "Build-in wiretap channel I with feedback and LDPC codes," J. Commun. Netw., vol. 11, no. 6, pp. 538–643, Dec. 2009.
27. Hong Wen, Pin-Han Ho and Xiao-Hong Jiang, "On achieving unconditional secure communications over binary symmetric channels (BSC)," IEEE Wirel. Commun. Lett., vol. 1, no. 2, pp. 49–52, 2012.
28. S. Lin and D. J. Costello, Jr., Error control coding: fundamentals and applications. Englewood Cliffs, NJ: Prentice-Hall, 1983.
29. H. Koorapaty, A. A. Hassan, S. Chennakeshu, "Secure information transmission for mobile radio," IEEE Trans. Wirel. Commun., vol. 2, no. 7, pp. 52–55, 2003.
30. Y. Zhang and H. Dai, "A real orthogonal space-time coded UWB scheme for wireless secure communications," EURASIP J. Wirel Commun. Netw., vol. 6, no. 3, pp. 1–8, 2009.
31. Y. Hua, S. An and Y. Xiang, "Blind identification of FIR MIMO channels by decorrelation subchannels," IEEE Trans. Signal Process., vol. 51, no. 5, pp. 1143–1155, 2003.
32. T. Liu, S. Shamai Shitz, "A note on the secrecy capacity of the multiantenna wiretap channel," IEEE Trans. Inf. Theory, vol. 55, no. 6, pp. 2547–2553, 2009.
33. Y. Nawaz and G. Gong, "WG: A family of stream ciphers with designed randomness properties," Inf. Sci., vol. 178, no. 7, pp. 1903–1916, 2008.
34. N. Courtois, "Higher order correlation attacks, XL algorithm and cryptanalysis of toyocrypt", In Proceedings of ICISC 2002, LNCS: vol. 2587, Berlin: Springer, pp. 182–199, 2003.
35. V. Tarokh, H. Jafarkhani, A. R. Calderbank, "Space time block codes from orthogonal designs," IEEE Trans. Inf. Theory, vol. 45, no. 5, pp. 744–765, 1999.
36. IEEE P802.11n, Draft standard for information technology telecommunications and information exchange between systems local and metropolitan area networks specific requirements, Part 11:Wireless LAN Medium Access Control (MAC) and Physical Layer (PHY) specifications. 802.11 Working Group of the 802 Committee, 2009.
37. S. M. Alamouti, "A simple transmitter diversity scheme for wireless communications," IEEE Journal on Selected Areas in Communications, vol. 16, no.8, pp. 1451–1458, 1998.

38. V. Tarokh, A. Naguib, N. Seshadri, A. R. Calderbank, "Space-time codes for high data rate wireless communication: performance criteria in the presence of channel estimation errors, mobility, and multiple paths," IEEE Trans. Commun., vol. 17, no. 2, pp. 199–207, 1999.

39. H. Wen, G. Gong, S. Lv and P. Ho, "Framework for MIMO cross-layer secure communication based on STBC," Telecommun. Syst. J., pp. 1–9, August 2011.

40. H. Wen, P. Ho and G. Gong, "A framework of physical layer technique assisted authentication for vehicular communication networks," Sci China Ser F-Inf Sci, vol. 53, no. 10, pp. 1996–2004, 2010.

41. H. Wen, P. Ho, C. Qi and G. Gong, "Physical layer assisted authentication for distributed Ad-Hoc wireless sensor networks," IEEE Inf. Secur., vol. 4, issue 4, pp. 390–396, 2010.

42. H. Wen, P. Ho, "Physical layer technique to assist authentication based on PKI for vehicular xommunication networks," KSII Trans. Internet Inf. Syst.," vol. 5, issue 5, pp. 440–456, Feb. 2011.

43. H. Wen, J. Luo and L. Zhou, "Lightweight and effective detection scheme for node clone attack in WSNs," IET Wireless Sensor Systems, vol. 1, no. 3, pp. 137–143, Sept. 2011.

44. H. Wen, P. Ho and X. Jiang, "On achieving unconditional secure communications over binary symmetric channels (BSC)," IEEE Wirel. Commun. Lett., vol. 1, no. 2, pp.49–52, 2012.

45. L. Xiao, L. Greenstein, N. Mandayam, W. Trappe, "Using the physical layer for wireless authentication in time-variant channels," IEEE Trans. Wirel. Commun., vol. 7, issue 7, pp. 2571–2579, July 2008.

46. L. Xiao, L. Greenstein, N. Mandayam, W. Trappe, "Fingerprints in the ether: using the physical layer for wireless authentication," in Proceedings of IEEE International Conference on Communications, pp. 4646–4651, 24–28 June 2007.

47. L. Xiao, L. Greenstein, N. Mandayam, W. Trappe, "A physical-layer technique to enhance authentication for mobile terminals," in Proceedings of IEEE International Conference on Communications, pp. 1520–1524, 19–23 May 2008.

48. P. L. Yu, J. S. Baras, B. M. Sadler, "Physical-layer authentication," IEEE Trans. Inf. Forensics and Secur., vol. 3, no. 1, pp. 38–51, Mar. 2008.

49. P. A. Bello, "Characterization of randomly time-variant linear channels," IEEE Trans. Comm. Syst., vol. 11, pp. 360–393, 1963.

50. O. Edfors, M. Sandell, J. J. van de Beek, S. K. Wilson, and P. O. Borjesson, "OFDM xhannel estimation by singular value decomposition," IEEE Trans. Comm., vol. 46, no. 7, pp. 931–939, July 1998.

51. P. Hoeher, S. Kaiser, and P. Robertson, "Pilot-symbol-aided channel estimation in time and frequency," in Proceedings of IEEE Global Telecommunications, pp. 90–96, Nov. 1997.

52. Y. Li, L. J. Cimini, Jr. and N. R. Sollenberger, "Robust channel estimation for OFDM systems with rapid dispersive fading channels," IEEE Trans. Commun., vol. 46, no. 7, pp. 902–915, July 1998.

53. S. Coleri, M. Ergen, A. Puri, A. Bahai, "A Study of channel estimation in OFDM systems," in Proceedings of IEEE VTC, vol. 2, pp. 894–898, Vancouver, Canada, Sept. 2002.

54. Y. Qiao, S. Yu, P. Su, and L. Zhang, "Research on an iterative algorithm of LS Channel estimation in MIMO OFDM systems," IEEE Trans. Broadcast, vol. 51, no. 1, pp. 149–153, Mar. 2005.

55. A. Wald, "Sequential tests of statistical hypotheses," Ann. Math. Stat. 16 (2): 117–186, June 1945.

56. "Dedicated Short Range Communications (DSRC)," [Online] Available: http://grouper.ieee. org/ groups /scc32/dsrc/index.html, 2007.

57. Task Group p, "IEEE P802.11p: Draft standard for information technology telecommunications and information exchange between systems Local and metropolitan area networks Specific requirements, Part 11: Wireless LAN Medium Access Control (MAC) and Physical Layer (PHY) specifications," IEEE Computer Society, Jun. 2009.

58. J.P. Hubaux, "The security and privacy of smart vehicles," IEEE Secur. Priv., vol. 2, pp. 49–55, 2004.

59. M. Raya and J. P. Hubaux, "Securing vehicular ad hoc networks," J. Comput. Secur., vol. 15, no. 1, pp. 39–68, 2007.

60. F. Dotzer, "Privacy issues in vehicular Ad Hoc networks," in Proceedings of ACM Workshop on Vehicular Ad Hoc Networks, Sept. 2006.

61. H. Moustafa, G. Bourdon, and Y. Gourhant, "AAA in vehicular communication on highways with Ad Hoc networking support: a proposed architecture," in Proceedings of ACM workshop on Vehicular ad hoc networks, pp. 79–80, 2005.

62. C. Zhang, R. Lu, X. Lin, Pin-Han Ho, and X. Shen, "An efficient identity-based batch verification scheme for vehicular sensor networks," in Proceedings of IEEE INFOCOM, pp. 246–250, 2008.

63. X. Lin, X. Sun, X. Wang, C. Zhang, Pin-Han Ho, X. Shen, "TSVC: timed efficient and secure vehicular communications with privacy preserving," IEEE Trans. Wirel. Commun., vol. 7, no. 12, pp. 4987–4998, 2009.

64. C. Zhang, X. Lin, R. Lu and P. H. Ho, "RAISE: an efficient RSU-aided message authentication scheme in vehicular communication networks," in Proceedings of ICC'08, pp. 1451–1457, May 19–23, 2008.

65. Adrian Perrig, Robert Szewczyk, Victor Wen, David Culler, J. D. Tygar, "SPINS: security protocols for sensor netowrks," in Proceedings of the 7th annual international conference on Mobile computing and networking, July 2001, Rome, Italy, p. 189–199.

66. Donggang Liu and Peng Ning, "Multilevel μTESLA: broadcast authentication for distributed sensor networks," ACM Trans. Embed. Comput. Syst., vol. 3, no. 3, Nov. 2004, pp. 800–836.

67. M. Demirbas, Youngwhan Song, "An RSSI-based scheme for Sybil attack detection in wireless sensor networks," in Proceedings of WoWMoM 2006, pp. 566–570, 2006.

68. R. L. Rivest, A. Shamir, and L. M. Adleman, "A method for obtaining digital signatures and public-key cryptosystems," Commun. ACM, vol. 21, no. 2, 1978, pp. 120–126.

69. U. S. National Institute of Standards and Technology (NIST), Digital Signature Standard (DSS), Federal Register 56. FIPS PUB 186, Aug. 1991.

70. U. S. National Institute of Standards and Technology (NIST). DES Model of Operation. Federal Information Processing Standards Publication 81 (FIPS PUB 81).

71. S. P. Miller, C. Neuman, J. I. Schiller, and J. H. Saltzer, "Kerberos authentication and authorization system," In Project Athena Technical Plan, page section E.2.1, 1987.

72. IEEE LAN/MAN Standards Committee, "IEEE 802.16a: air interface for fixed broadband wireless access systems," 2003.

73. Wireless InSite software, http://www.remcom.com/wirelessinsite/.

74. J. Daemen and V. Rijmen. AES proposal: Rijndael, Mar. 1999.

75. IEEE P802 LAN/MAN Committee, "The working group for wireless local area networks (WLANs)," http://grouper.ieee.org/groups/802/11/index.html.

76. B. Parno, A. Perrig, and V. Gligor, "Distributed detection of node replication attacks in sensor networks," in Proceedings of the IEEE Symposium on Security and Privacy, pp. 49–63, 2005.

77. C. Karlof and D. Wagner, "Secure routing in wireless sensor networks: attacks and countermeasures," In Proceedings of First IEEE International Workshop on Sensor Network Protocols and Applications, pp. 113–127, 2003.

78. Anthony D. Wood and John A. Stankovic. A Taxonomy for denial-of-service attacks in wireless sensor networks. Handbook of Sensor Networks: Compact Wireless and Wired Sensing Systems, 2004.

79. Y. Zhang, W. Liu, W. Lou, and Y. Fang, "Location-based compromise-tolerant security mechanisms for wireless sensor networks," IEEE J. Sel. Areas Commun., vol. 24, no. 2, pp. 247–260, 2006.

80. R. Brooks, P. Y. Govindaraju, M. Pirretti, N. Vijaykrishnan, and M. T. Kandemir, "On the detection of clones in sensor networks using random key predistribution," IEEE Trans. Syst., Man, and Cybern., Part C: Applications and Reviews, vol. 37, no. 6, pp. 1246–1258, 2007.

81. C. Bekara and M. Laurent-Maknavicius, "A new protocol for securing wireless sensor networks against nodes replication attacks," in Proceedings of Third IEEE International Conference on Wireless and Mobile Computing, Netw. Commun. (WiMOB 2007), 2007, pp. 59–59.

82. Zhijun Li and Guang Gong, "DHT-based detection of node clone in wireless sensor networks," in Proceedings of First International Conference on Ad Hoc Networks (ADHOCNETS 2009), Sept. 23–25, 2009, Niagara Falls, Ontario, Canada, LNICST 28, pp. 240–255.

83. Zhijun Li and Guang Gong, "Randomly directed exploration: An efficient node clone detection protocol in wireless sensor networks," in Proceedings of IEEE 6th International Conference on Mobile Adhoc and Sensor Systems (MASS '09), Oct. 12–15, 2009, Macau SAR, P.R.C, pp. 1030–1035.

84. T. Suen and A. Yasinsac, "Ad Hoc network security: peer identification and authentication using signal properties," In Systems, Man and Cybernetics (SMC) Information Assurance Workshop, pp. 432–433, 2005.

85. Y. Sheng, K. Tan, G. Chen, D. Kotz, and A. Campbell, "Detecting 802.11 Mac layer spoofing using received signal strength," In Proceedings of the IEEE International Conference on Computer Communications (INFOCOM), Apr. 2008.

86. J. Douceur, "The Sybil attack," In Proceedings of First International Workshop on Peer-to-Peer Systems, pp. 251–260, 2002.

87. Zhang Jian-Ming, Yu Qun and Wang Liang-Min, "Geographical location-based scheme for Sybil attacks detection in wireless sensor networks," J. Syst. Simul., vol. 20, no. 1, pp. 259–263, Jan. 2008.

88. J. Newsome, E. Shi, D. Song, A. Perrig, "The Sybil attack in sensor networks: analysis & defenses," In Proceedings of third International Symposium on Information Processing in Sensor Networks, IPSN 2004, pp. 259–268, 2004.

89. L. Eschenauer and V. D. Gligor, "A key-management scheme for distributed sensor networks," In Proceedings of the 9th ACM Conference on Computer and communications security (CCS), Nov. 2002.

90. P. Golle, D. Greene and J. Staddon, "Detecting and correcting malicious data in VANETs," In proceedings of 1st ACM Workshop on Vehicular Ad Hoc Networks pp. 29–37, 2004.

91. Jie Yang, Yingying Chen, W. Trappe, "Detecting Sybil attacks in wireless and sensor networks using cluster analysis," In Proceedings of 5th IEEE International Conference on Mobile Ad Hoc and Sensor Systems, MASS 2008, pp. 834–839, 2008.

92. Ren Xiu-li, Yang Wei, "Method of detecting the Sybil attack based on ranging in wireless sensor network," in Proceedings of WiCom '09, pp. 1–4, 2009.